Calcium and Cellular Metabolism

Transport and Regulation

Calcium and Cellular Metabolism

Transport and Regulation

Edited by

J. R. Sotelo and
J. C. Benech

Instituto de Investigaciones Biológicas Clemente Estable
Montevideo, Uruguay

Plenum Press • New York and London

Library of Congress Cataloging in Publication Data

Calcium and cellular metabolism: transport and regulation / edited by J. R. Sotelo and
J. C. Benech.
      p.   cm.
    "Proceedings of an International Workshop on Calcium and Cellular Metabolism:
Transport and Regulation, held September 25–October 6, 1995, in Montevideo, Uru-
guay"—T.p. verso.
    Includes bibliographical references and index.
    ISBN 0-306-45594-3
    1. Calcium channels—Congresses. 2. Potassium channels—Congresses. 3. Calcium—
Physiological effect—Congresses. I. Sotelo, J. R. II. Benech, J. C. III. International
Workshop on Calcium and Cellular Metabolism: Transport and Regulation (1995: Mon-
tevideo, Uruguay)
QP535.C2C6185   1997   97-12021
571.6′4—dc21   CIP

Proceedings of an International Workshop on Calcium and Cellular Metabolism: Transport and
Regulation, held September 25 – October 6, 1995, in Montevideo, Uruguay

ISBN 0-306-45594-3

© 1997 Plenum Press, New York
A Division of Plenum Publishing Corporation
233 Spring Street, New York, N. Y. 10013

http://www.plenum.com

10 9 8 7 6 5 4 3 2 1

Printed in the United States of America

PREFACE

When we began to organize the workshop "Calcium and Cellular Metabolism: Transport and Regulation" the goal we had in mind was to put together the knowledge of several specialists on Ca^{2+} homeostasis, with various examples of cellular metabolisms (such as protein synthesis), regulated by Ca^{2+} ions. Regarding the homeostasis of Ca^{2+} ions, we invited Ernesto Carafoli to write the first chapter as a general state-of-the-art introductory review. On the other hand, the other chapters are the contribution of different specialists on membrane calcium transport mechanisms, aiming to reunite at least in part the wide field of calcium homeostasis. We roughly try to group chapters that share similar subjects.

The first group of chapters (Chapters 2 to 6), are mainly related to calcium channels. Thus, Chapter 2 by Rodolfo Llinás et al. describes a new concept related to the dimensions of the calcium action domain at the inner mouth of calcium channels in the squid giant synapse and its relationship to neurotransmitter release. Chapter 3 by Martin Morad et al. informs us about new ways of identifying and measuring, by confocal microscopy, individual sites where calcium release occurs in ventricular myocytes. In the same group Osvaldo Uchitel and Eleonora Katz classify and evaluate the variety of calcium channels at the neuromuscular junction, in Chapter 4. Chapter 5 by Gustavo Brum et al. compares the currents through L-type calcium channels and charge movements in cardiac and skeletal muscle, in relation to voltage dependent inactivation of calcium channels. Finally, Chapter 6 by Ramón Latorre et al., although not directly related to calcium channels, describes the inactivation of the *Shaker* potassium channel which has the peculiarity of sharing the S4 intramembrane domain with calcium and sodium channels, grouping them in the so-called S4 superfamily.

The second group of chapters (Chapters 7, 8, and 9) relates to calcium pumps. Chapter 7 by E. Carafoli and Danilo Guerini describes the plasma membrane calcium pump (PMCA), its relationship to calmodulin and its targeting to the plasma membrane. Chapter 7 comes within the third and final group of Chapters described later. Chapter 8 by Leopoldo de Meis describes the peculiar thermodynamics of the $Ca^{2+}ATPase$ responsible for calcium transport into the endoplasmic reticulum, beautifully inserted in its historical frame. Finally, Chapter 9 by Luis Beaugé et al. is related to the dephosphorilation steps of $Na^+K^+ATPase$, emphasizing the idea that the free energy stored in the sodium electrochemical gradient is used to extrude calcium through the Na^+-Ca^{2+}-exchanger.

The third group of chapters (Chapters 10 to 13) shows some examples of calcium regulation of cell metabolism. Chapter 10 of Vivian Rumjanek et al. describes the Multi Drug Resistance (MDR) phenomenom, a major reason for failure in human chemotherapy, mediated by a glyco-protein transporter belonging to the family of traffic ATPases, and its relation to calcium homeostasis. Finally, Chapters 11, 12, and 13 are related to neuronal protein synthesis regulation by calcium. In this regard we consider that Chapter 7 of E. Carafoli and D. Guerini may be included in this group. This chapter reports the dependence of PMCA expression upon changes on intracellular calcium concentration following cell depolarization. Cell depolarization was induced by extracellular high potassium concentrations. In Chapter 11, J. R. Sotelo et al. describe some experiments regarding the so-called local protein synthesis in the axonal territory, and how this local protein synthesis and the perikaryal protein synthesis are regulated by calcium ions. In Chapter 12, Antonio Giuditta (who together with Edward Koenig have been pioneers on the study of protein synthesis in invertebrate and vertebrate axons, respectively), presents an exhaustive review on the most relevant findings regarding protein synthesis in the squid giant axon and presynaptic terminals of the squid brain. In Chapter 13, J. C. Benech et al. describe how protein synthesis in the squid brain presynaptic territory is regulated by Ca^{2+} ions. We attribute great relevance to the regulation of protein synthesis by calcium in neurons giving the well-known hypothesis of calcium as a second messenger involved in the communication of external cell signals (transmitted by excitable cell membranes) or intracellular signals, and gene translation (protein synthesis). We strongly believe that in a near future these functions of Ca^{2+} ions may be important to explain the transfer of information from short-term neuronal functions to long-term functions, such as neuronal plasticity, memory, or learning, as well as the regulation of other functions like nerve regeneration and nerve cell development.

The editors are indebted to all the scientists who have contributed to this book and especially helped us in improving our work. We particularly want to mention Dr. Patricio Garrahan and Dr. Carlos Gutierrez Merino, who participated in the Workshop but could not contribute to the book for personal reasons. The editors also want to acknowledge the financial support of: the International Union of Pure and Applied Biophysics (IUPAB), International Union of Biochemistry and Molecular Biology (IUBMB), Programa de Desarrollo de Ciencias Básicas (PEDECIBA), Ministerio de Educación y Cultura (MEYC), Programa BID-CONICYT, and Projects: CT1-CT93–0037UY of the European Union, Proyecto 058, BID-CONICYT and Proyecto Diabetic Neuropathy, Japanese International Cooperation Agency (JICA).

We also want to acknowledge the Instituto de Investigaciones Biológicas Clemente Estable (IIBCE, MEYC), the School of Sciences (University of the Republic, U.R.) and the School of Medicine (U.R.) for generously providing all the necessary facilities for developing this workshop. Finally, special thanks to Adela Wittenberger and Laura Diaz Arnesto, Ph.D., for their valuable language advice as well as to all the generous people that helped us individually.

CONTENTS

REGULATION OF CALCIUM SIGNALLING IN CELLS

Ernesto Carafoli

Institute of Biochemistry
Swiss Federal Institute of Technology (ETH)
CH-8092 Zürich, Switzerland

1. INTRODUCTION

Ca controls numerous cellular functions (see Carafoli, 1987, for a comprehensive review). Its signalling role demands its maintenance within cells at a very low free concentration; thus, mechanisms exist to modulate it in the cell domains where the Ca-sensitive targets are located. To achieve this, evolution has selected the reversible complexation by proteins, which are either soluble, organized in non-membranous structures, or intrinsic to membranes: they complex intracellular Ca from concentrations which are about 10,000 fold lower than in the external spaces. The Ca controlling function of the plasma membrane is based on the operation of membrane-intrinsic Ca transporting proteins, and is responsible for the long-term maintenance of the Ca gradient between cells and the extracellular space. The large inwardly directed Ca gradient maintained by the plasma membrane transporters ensures that even minor changes in the Ca permeability of the plasma membrane will produce significant swings in its intracellular concentration, and thus in turn efficiently modulate its signalling function.

2. THE HIGH AFFINITY INTRACELLULAR Ca BINDING PROTEINS

The crystal structure of the Ca-binding protein parvalbumin led Kretsinger and his associates to derive a set of rules for the binding of Ca by high affinity Ca binding proteins, these proteins are now known as the EF-hand proteins, which form a family with more than 200 members (Kretsinger & Nockoleds, 1973). The rules predict domains made of two orthogonal α-helices, interrupted by a loop of 10–12 amino acids where Ca coordinates to oxygen atoms (Crumpton & Dedman, 1990; Barton et al., 1991; Huber et al., 1990) of carboxylic side chains and to carbonyl or water oxygens. This helix-loop-helix motif has been optimised in the course of evolution and has been shown to be contained even in some extracellular proteins. It

Calcium and Cellular Metabolism: Transport and Regulation, edited by Sotelo and Benech.
Plenum Press, New York, 1997

has been confirmed by the solution of some crystal structures (see, for example Szebenyi et al., 1981; Herzberg & James, 1985; Babu et al., 1985).

Other intracellular Ca-binding proteins, the annexins, have also attracted considerable attention. They bind to negatively charged cell membranes in a Ca-dependent manner (Crumpton & Dedman, 1990), and may play a role in the Ca-modulated interactions of the cytoskeleton and in membrane fusion. They are also inhibitors of phospholipase. They contain repeats of about 75 amino acids separated by sequences of variable length (Barton et al., 1991). The crystal structure of some annexins, e.g., that of annexin V (Huber et al., 1990; Concha et al., 1993; Weng et al., 1993; Huber, Schneider et al., 1990, Bewley et al. 1993), shows five α-helices bundled into a right-handed super-helix, leading to the suggestion that they could form a trans-membrane calcium channel (Huber, Schneider et al. 1990). The Ca coordinating residues in the annexins are not adjacent in the sequence: of the three calcium binding sites, two are located in repeats II and IV, one in either repeat I or III (Concha et al., 1993; Weng et al., 1993; Bewley et al., 1993). Also in the case of annexins the Ca atoms are coordinated to 7 oxygens from peptide carbonyls or water molecules.

Although the soluble EF-hand Ca-binding proteins contribute to the buffering of cell Ca^{2+} their main function is to "decode" the Ca information: they do so by exposing hydrophobic domains upon binding Ca: this enables them to interact with (protein) targets. A second conformational change is the collapse of the elongated structure of the protein around its binding domain in the target enzymes, which completes the processing of the Ca signal. The solution of the tridimensional NMR and crystal structures of the complex of calmodulin, with its binding domain in two target proteins has documented the conformational change (Ikura et al., 1992; Meador et al., 1992). Figure 1 summarises the NMR

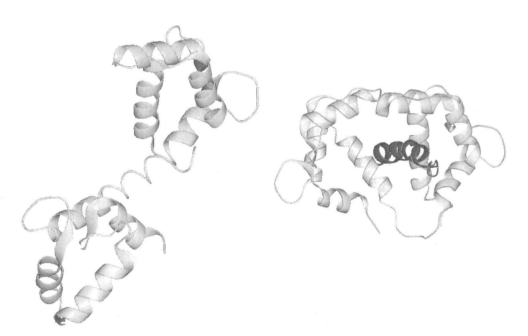

Figure 1. The tertiary (NMR) structure of calmodulin complexed with its synthetic binding domain in myosin light chain kinase (Ikura et al., 1992). Left, the structure of uncomplexed, extended calmodulin. Right: calmodulin bends around its binding domain in myosin-like chain kinase (the darker peptide).

results, in which a synthetic calmodulin binding peptide corresponding to the calmodulin binding domain of myosin light chain kinase has been used.

3. MEMBRANE TRANSPORT OF CALCIUM

The Ca^{2+} buffering function of EF-hand proteins (and, possibly, annexins) is quantitatively limited by their total amount in cells: the total Ca buffering capacity of the two most important EF-hand proteins, i.e., calmodulin and troponin C has been estimated at 6–40 μM Ca for calmodulin in most tissues, and at 8 and 80 μM Ca in heart and skeletal muscles, respectively.

The quantitative limitations in the Ca buffering function of the soluble proteins do not apply to membrane intrinsic Ca-binding proteins, which complex Ca at one membrane side, transport it across and repeat the operation continuously. These proteins thus play the most important role in the buffering of intracellular Ca. Several are present in eucaryotic cells, both in the plasma membrane and in the organelles (Figure 2). They satisfy both the demand for rapid, precise, and high affinity regulation, or for lower affinity and less rapid regulation. These proteins function according to four basic transporting mechanisms: ATPases, exchangers, channels, electrophoretic uniporters. High affinity Ca regulation depends obligatorily on ATPases, whereas lower affinity regulation uses any of the other transport modes.

3.1. Ca Transport in the Plasma Membrane

Three Ca transporting systems operate in the plasma membrane: Ca channels, a Na/Ca exchanger and a Ca pump. The Ca channels are responsible for the penetration of Ca: although they can in some cases be gated by the interaction with ligands, in most cases they are sensitive to voltage. They are subdivided in two sub-classes: the T (tiny) channels, which are activated and inactivated at low membrane potentials, and the channels which are activated and inactivated at high voltage. The latter have been sub-divided into four classes: the B (brain) channels, the L (long-lasting) channels, the N (neither L

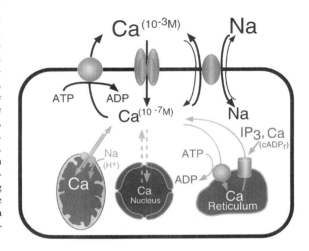

Figure 2. Ca transporting systems in the membranes of eucaryotic cells. The systems in the nuclear envelope are less well characterised, and are thus represented by dashed lines. Three systems are known in the plasma membrane (a Ca-ATPase, a Na/Ca exchanger, which normally exports Ca, but sometimes also imports it, and a Ca channel, of which several types have been described, see text); calcium channels are also active in the endo(sarco)plasmic reticulum, but they are gated by ligands, not by voltage. The reticulum also contains a calcium ATPase. Two calcium transporting systems are present in the inner membrane of mitochondria (an electrophoretic uptake uniporter and a Ca releasing Na/Ca exchanger. The inner membrane of some mitochondrial types, however, contains a Ca/proton exchanger). (IP3, inositol 1, 4, 5-triphosphate; cADPr, cyclic ADP ribose).

nor B) channels, and the P (Purkinye cells) channels. The best known are the L-channels (Reuter, 1984), which have a conductance of 15–25 pS, corresponding to the passage of about 3 x 10^6 Ca ions per sec (Reuter et al., 1982). The L-channels begin to open as the transmembrane potential increases to about -40 mv (maximal Ca currents are seen at about 0 mv), and are blocked by several classes of Ca antagonists (Fleckenstein, 1973). They are composed of 5 subunits: $\alpha 1$ (which contains the trans-protein Ca channel), $\alpha 2$, β, γ, δ. The $\alpha 2$ and δ subunits are linked by disulphide bonds. The γ and δ-subunits are membrane intrinsic, while the β subunit is located at the inner side of the plasma membrane. The primary structure of all subunits is now known (see Hofmann et al., 1994 for a recent review) and the sequence of the channel-forming $\alpha 1$ subunit repeats the architectural motifs of other voltage-gated membrane channels, i.e., 4 repeats of 6 transmembrane domains each. The fourth transmembrane domains in each repeat contain several positively charged amino acids and is assumed to be the voltage sensor of the channel. The latter is assumed to be formed by the "external" loop connecting transmembrane domains 5 and 6 in each of the 4 repeats, which would fold within the membrane environment to provide the path for Ca (Figure 3 adapted from Hofmann et al., 1994). The role of the other L-channel subunits is not yet clear, except that the β-subunit is phosphorylated by protein kinase A, and is thus probably responsible for the increased probability of L-channel opening induced by β-adrenergic stimulation of cells, e.g., heart cells.

The Na/Ca exchanger is one of the two Ca exporting systems of the plasma membrane but under some conditions can also mediate the influx of Ca. It is a large capacity, low affinity system, particularly active in excitable tissues like heart. Electrophysiological work has established that it operates electrogenically, exchanging 3 Na for 1 Ca, and thus responds also to the transmembrane electrical potential. The system has low affinity for Ca (K_m, 1–20µM), but a high rate of transport, (in heart, about 20 nmoles of Ca per mg of sarcolemmal protein per sec.). The low Ca affinity of the system would, in principle, be difficult to reconcile with the function of the exchange system in the export of Ca from cells. However, it is assumed that the exchanger is distributed inhomogeneously in the plasma membrane, and is thus specifically exposed to areas of localized high Ca concentration. This is the case, for instance, in heart, where the exchanger is preferentially located in the T-system portion of the plasma membrane, and would thus have immediate access to the high concentrations of Ca produced by the operation of the Ca-release channels of the terminal cisternae of sarcoplasmic reticulum.

The exchanger (Nicoll et al., 1990) has been shown to be organized in the sarcolemma with 11 transmembrane domains (Figure 4). A large cytosolic loop separates transmembrane domains 5 and 6, and contains a site for Ca binding and a putative calmodulin binding site (putative because no evidence is available for the regulation of the exchanger by calmodulin in vivo). The loop is the locus of particularly complex alternative mRNA splicing events which produce a large number of potential isoforms (Kofuii et al., 1994). Surprisingly, however, the loop can be deleted in heterologous expression experiments without total loss of exchanger activity (Matsuoka et al., 1993). A variant of the exchanger, which is the product of a separate gene and is typical of brain, is also known. Although the importance of the exchanger in the control of Ca ejection is established in excitable tissues, in non-excitable tissues the exchanger may play a comparatively minor role. A particular variant of the exchanger is active in the plasma membrane of retinal cones and rods: although its membrane topography seems to be similar to that of the heart-type exchanger (Reiländer et al., 1992), the retinal protein is mechanistically and structurally different. It has no particular sequence homology to the heart protein and transports K as well as Na and Ca with a stoichiometry of 4 sodium ions against one calcium ion and one potassium ion.

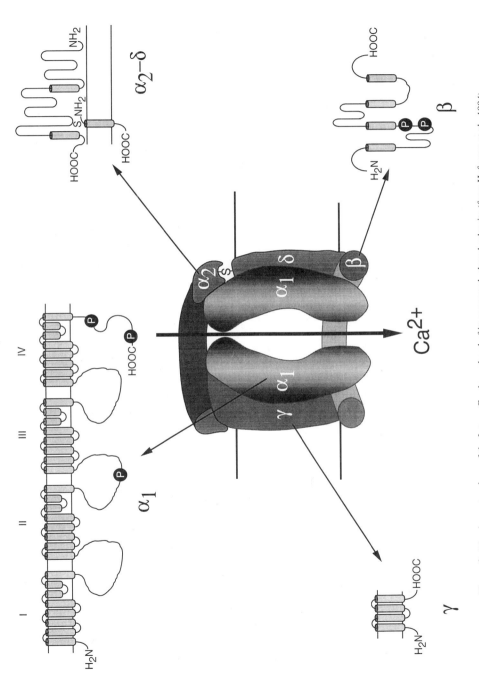

Figure 3. Membrane topology of the L-type Ca channel and of its separated, cloned subunits (from Hofmann et al., 1994).

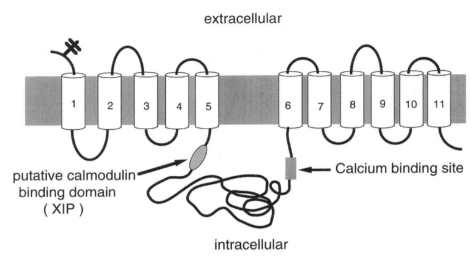

Figure 4. Membrane topology and location of some functional domains of the heart Na/Ca exchanger (from Nicoll et al., 1990). The calmodulin binding domain (a peptide termed XIP) is putative, because no calmodulin modulation of the exchanger in vivo has been so far documented.

The Ca ATPase (pump) of the plasma membrane interacts with Ca with high affinity (K_m, 0.5 µM or less), but has low transport capacity: (about 0.5 nmol per mg of membrane protein per sec). The enzyme thus exports Ca from cells even when its concentration in the cytosol is at the normal sub-µMolar level, and plays the most important role in maintaining the gradient of Ca between cells and medium in most eucaryotic cells. The enzyme belongs to the P-class of transport ATPases, i.e., it forms an aspartyl phosphate during the reaction mechanism and is inhibited by vanadate (see Carafoli, 1991; Carafoli, 1992 for recent reviews). It is stimulated by calmodulin and transports Ca with a 1:1 stoichiometry to ATP hydrolysis, as compared to the stoichiometry of 2 for the analogous ATPase of sarcoplasmic reticulum (see below). A cAMP-linked phosphorylation process stimulates the ATPase and the associated pumping of Ca in one of the isoforms of the pump. Protein kinase C also stimulates the pump, although to a lesser extent. Treatments alternative to calmodulin and to the kinase-mediated phosphorylation, e.g., the exposure to acidic phospholipids and polyunsaturated fatty acids, a controlled proteolysis by a number of proteases (e.g., calpain) and a dimerisation (oligomerization) process also stimulate the pump.

The primary structure of the pump (Shull & Greeb, 1988; Verma et al, 1988) shows in 10 transmembrane domains and most of the mass of the pump in the cytoplasm (Figure 5). The largest cytoplasmic protruding unit contains the active site, whereas the unit protruding from the 10th transmembrane domain contains the calmodulin binding site and the substrate sites for protein kinases. The calmodulin binding domain interacts with the pump at two sites located in the first and second cytoplasmic protruding domains: this makes the access to the active site more difficult, and explains the inhibited state of the pump in the absence of calmodulin. Calmodulin interacts with its own binding domain and removes the latter from the binding site in the pump, relieving the inhibition. Thus, this activation/deactivation mechanism of the plasma membrane ATPase closely resembles that of the sarcoplasmic reticulum Ca pump (see below) where the accessory membrane protein phospholamban binds next to the active site maintaining it inhibited. Removal of phospho-

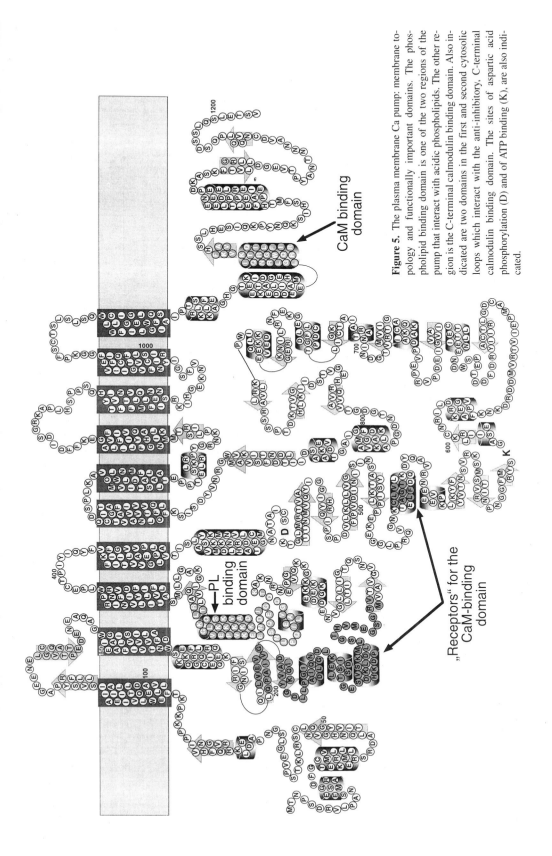

Figure 5. The plasma membrane Ca pump: membrane topology and functionally important domains. The phospholipid binding domain is one of the two regions of the pump that interact with acidic phospholipids. The other region is the C-terminal calmodulin binding domain. Also indicated are two domains in the first and second cytosolic loops which interact with the anti-inhibitory, C-terminal calmodulin binding domain. The sites of aspartic acid phosphorylation (D) and of ATP binding (K), are also indicated.

CaM binding domain

PL binding domain

"Receptors" for the CaM-binding domain

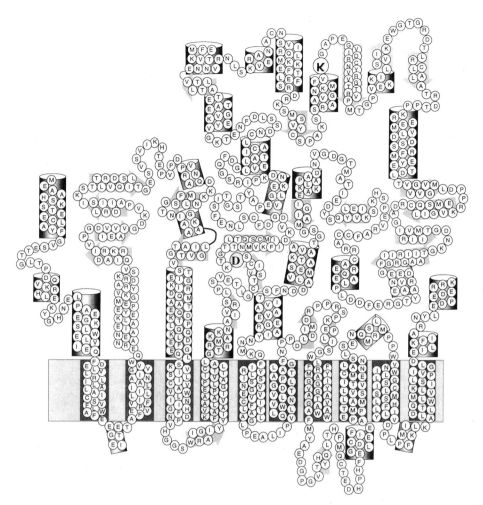

Figure 6. The calcium pump of the sarcoplasmic reticulum. The sequence similarity to the calcium pump of the plasma membrane is limited, but the membrane architecture of the two pumps is very similar. The reticular pump lacks the C-terminal protruding unit that contains the calmodulin binding domain. The sites of aspartic acid phosphorylation (D) and of ATP binding (K) are also indicated.

lamban from its site of interaction, however, is accomplished by its phosphorylation by two kinases, not by calmodulin binding.

The homology between the two modulation mechanisms would thus be incomplete, if recent experiments on the plasma membrane Ca pump had shown (Hofmann, Anagli et al, 1994) that the phosphorylation of the calmodulin binding domain by protein kinase C also removes its ability to interact with its receptor sites in the pump. Hence, the regulation mechanism of the two pumps are strikingly similar.

In humans the plasma membrane Ca pump is the product of a multigene family. Of the four genes, two (genes 1 and 4) are expressed in large and approximately similar amounts in all tissues studied (Stauffer et al, 1993). The two other genes are expressed in a tissue-specific way, brain being the tissue where their products are most conspicuously represented. Additional isoform diversity is produced by the alternative splicing of the primary transcripts of all four genes. The distribution of the spliced variants of the pump is also tissue-specific: once again, brain is the tissue where the largest number of variants is found.

3.2. Ca Transport across the Membranes of Cell Organelles

Although the long-term maintenance of the Ca gradient between cells and medium depends on the operation of the plasma membrane transport systems, in most cells the plasma membrane fluxes are minor compared to the total amount of Ca involved in the functional cycle of the cell. Although Ca crossing the plasma membrane triggers important intracellular events and thus is vital to cell function, most of the Ca needed for cell activity is extracted from intracellular stores.

3.2.1. The Endo(sarco)plasmic Reticulum. Most of the early work on this membrane system has been performed on sarcoplasmic reticulum of muscle and on its Ca pump which is similar in mechanism and membrane topography to that of the plasma membrane. Recently, endoplasmic reticulum has also been studied intensively, due to the discovery that its Ca pool is sensitive to important second messengers, e.g., inositol-trisphosphate (Streb et al., 1983; Berridge, 1993).

The Ca-release channels in the reticulum system of muscle and non-muscle cells (see Berridge, 1993 for a recent comprehensive review), now include several types: one is the ryanodine sensitive channel of muscles, which is Ca-modulated (Fabiato & Fabiato, 1975) and is thus responsible for the phenomenon of Ca-induced Ca-release, generally assumed to mediate excitation-contraction coupling. Another, recently discovered channel is that sensitive to cyclic ADP-ribose (Galione, 1993) which may only be active in some cell types. No structural information is as yet available on the latter channel, but both the inositol-tris-phosphate (Furuichi et al., 1989) and the ryanodine-sensitive (Takeshima et al., 1989) channels have been cloned. They are very large oligomeric molecules which have analogies in membrane topology, i.e., both have an even number of transmembrane domains (4 in the ryanodine sensitive channel at least in the original cloning assignments, 6 in that sensitive to IP_3) and large masses protruding into the cytosol. Both are regulated by protein kinases and contain calmodulin binding domains.

The Ca^{2+} ATPase of the endo(sarco)plasmic reticulum is similar to that of the plasma membrane, and has high affinity for Ca (K_m below 0.5 μM). It is very abundant in the membranes of sarcoplasmic reticulum, which thus have a large total Ca transporting capacity (in fast skeletal muscles it may reach 70 nmol per mg of membrane protein per sec). The ATPase has been purified long ago (MacLennan, 1970) as a single polypeptide of about 100 kDa: it forms an aspartyl-phosphate like the similar pump of the plasma membrane, is inhibited by vanadate and can be reconstituted into liposomes where it transports Ca with a 1:1 stoichiometry to the hydrolyzed ATP. Its primary structure, now determined (MacLennan et al., 1985) for both heart and fast skeletal muscles reticulum, repeats the essential topography motifs of the plasma membrane Ca pump, except for the absence of the C-terminal protruding units containing the calmodulin binding domain. As mentioned above, the ATPase (in heart, smooth and slow, but not fast skeletal muscles) is modulated by the acidic proteolipid phospholamban (Tada et al., 1975), a pentamer of five identical subunits of M_r about 6 kDa, which is phosphorylated by both the cAMP and the calmodulin dependent protein kinases. Although isolated phospholamban forms pentamers, it has not yet been conclusively established that the

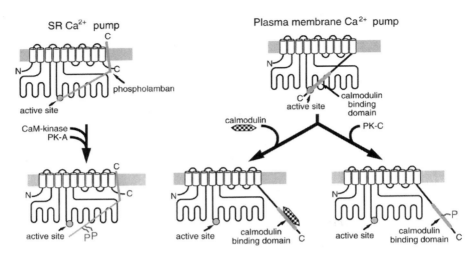

Figure 7. Analogies in the regulation of the plasma membrane and sarcoplasmic reticulum (SR) calcium pumps. In the plasma membrane pump, the interaction between calmodulin and its own binding domain removes it from its "receptor" site in the central portion of the pump, thus activating it. This resembles the mechanism in the sarcoplasmic reticulum pump where phospholamban, which has homology with the calmodulin-binding domain, binds close to the active site, inhibiting the pump. Removal of phospholamban from its site of interaction occurs through phophorylation by two kinases. The similarity of the activation/deactivation mechanisms of the pumps is made more striking by the finding that the calmodulin-binding domain can also be removed from its binding site in the plasma membrane pump by phosphorylation by protein kinase C, (PK-C, lower right) N, amino terminal; C, carboxyl E terminal.

pentamer is the functionally active unit in vivo. Phospholamban binds next to the active site of the Ca pump in the unphosphorylated state (James et al., 1989) and phosphorylation by the two kinases removes it from its binding site, activating the pump. It has also been claimed that a third kinase (protein kinase C) would phosphorylate phospholamban, thus explaining the smooth muscle relaxing effect of c-GMP (Cornwell et al., 1991). Figure 7 illustrates the homology in regulation mechanisms between the sarcoplasmic reticulum and plasma membrane Ca pumps.

3.2.2. Mitochondria. Mitochondria were initially considered as important in the regulation of cytosolic Ca (see Carafoli, 1982 for an early review), but the demonstration that they only handle Ca with low affinity (Crompton et al., 1976) and the finding that they contain much less Ca in situ than generally assumed have led to a revaluation of their importance as cytosolic Ca regulators: this function is now considered of minor importance. The main task of the mitochondrial Ca transporting system is now considered to be the regulation of matrix Ca, an important role due to the existence of 3 matrix dehydrogenases which are precisely controlled by Ca (Figure 8)—should this control fail, the delivery of reduction equivalents to the respiratory chain, and thus in turn the synthesis of ATP, would become seriously deranged.

Since a number of recent reviews on the more modern aspects of mitochondrial Ca transport have appeared (see, for example, Richter, 1992), it will only be briefly mentioned that mitochondria accumulate Ca through an electrophoretic route which is energised by the electrical component of the proton-motive force across the inner membrane and is inhibited by μM concentrations of ruthenium red. The route has low Ca affinity

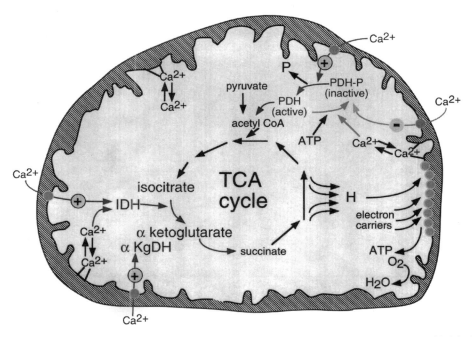

Figure 8. The regulation of three mitochondrial matrix dehydrogenases by Ca. PDH = pyruvic acid dehydrogenase; IDH = NAD-linked isocitric acid dehydrogenase; α-Kg-DH = α-ketoglutaric acid dehydrogenase.

(K_m, 1–10 μM) and accumulates Ca at a rate of up to 10 nmoles per mg of protein per sec in vitro. In vivo, however, the sub-optimal concentration of Ca in the cytosol and the presence of Mg, which also inhibits the route, reduce the uptake rate to a fraction of the optimal. Mitochondria also accumulate inorganic phosphate to precipitate Ca as an insoluble salt, probably hydroxyapatite, in the matrix. This results in the damping of the changes in the ionic Ca concentration in the latter compartment, limiting the disturbance to the Ca-modulated dehydrogenases by accumulated Ca.

Since the mitochondrial Ca uptake process is electrophoretic, and since the electrical potential across the inner membrane is unlikely to fluctuate, the release of Ca does not occur through the reversal of the uptake route, but through an electroneutral Na/Ca exchanger (Carafoli et al., 1974), which is particularly active in heart and other excitable tissues and which releases Ca at a slow rate (about 0.2 - 0.3 nmoles per mg of mitochondrial protein per sec). The Na-promoted release route, which is insensitive to ruthenium red, is blocked by some of the inhibitors of the plasma membrane Ca channel, e.g., the benzothiazepine diltiazem (Vaghy et al., 1982). Other mitochondrial types release Ca through a Ca/proton antiporter rather than through the Na/Ca exchanger (Richter, 1992). The energy-driven uptake route and the Na/Ca (or proton/Ca) exchanger can be integrated into an energy-dissipating Ca cycle (Carafoli, 1979, Figure 9): the low overall activity of the mitochondrial Ca transport systems *in vivo* limits the level of energy dissipation by the cycle.

The minor role of mitochondria in the regulation of cytosolic Ca under physiological conditions is thus essentially determined by the insufficient concentration of Ca in the cytosol. However, injuring conditions may damage the plasma membrane, permitting the en-

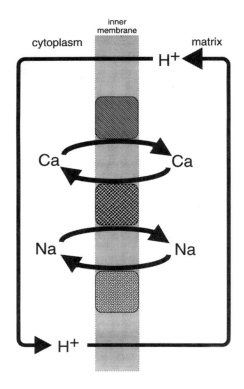

Figure 9. The energy dissipating Ca cycle of mitochondria. The cycle consists of the operations of three systems: an electrophoretic uptake uniporter, an electroneutral Na-Ca exchanger (in some mitochondrial types a Ca / proton exchanger is preferred) and an electroneutral Na-proton exchanger which returns excess Na to the cytosol.

try of excess Ca. If this occurs, the mitochondrial uptake system would become activated, leading to the storage of large amounts of precipitated Ca-phosphate in the matrix: these precipitates have indeed been repeatedly documented in electron microscopy observations of mitochondria of injured cells, including heart cells. If the injuring condition disappears, mitochondria slowly release the excess Ca at a rate compatible with the exporting ability of the plasma membrane transporting systems. Thus, they play a decisive protective role against the cytosolic Ca overload.

3.2.3. The Nucleus. The nucleus has so far attracted comparatively less attention as a partner in the process of calcium signalling: the general assumption that the nuclear pores are passively permeable to calcium has limited the interest on the control of nuclear calcium. The situation, however, is now changing, since several lines of evidence indicate that the calcium permeability of the nuclear pores may be controlled: (see Santella, 1996 for a recent review). An important factor in the renewed interest in the regulation of nuclear calcium has been the discovery that the nuclear envelope contains both a calcium ATPase (identical to that of the endoplasmic reticulum) and calcium channels modulated by IP3 and cADPr (Figure 10): clearly, these systems would be redundant if the nuclear pores were continuously and freely permeable to calcium. The concept that the concentration of calcium in the nucleoplasm must be carefully controlled is strongly supported by the discovery that the nucleus contains numerous reactions which are specific to it, and which are precisely modulated by calcium. A striking example is the transcription of some genes (Rosen et al., 1995), but essential steps in the process of programmed cell death could also be quoted in this connection.

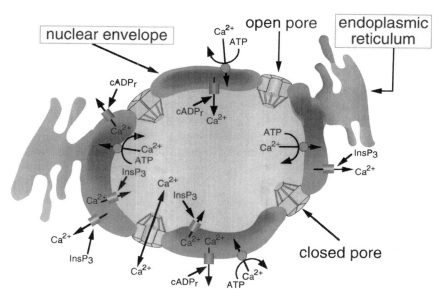

Figure 10. A scheme of the mechanisms for the control of calcium in the nucleus. The nuclear pores are represented in both the closed and permeable configurations.

4. CONCLUSIONS

The control of cell Ca is essentially performed by reversible complexation to specific proteins. Although soluble proteins contribute to Ca buffering, they are more important in the transmission of the Ca signal to targets. Membrane-intrinsic proteins play the main role in the buffering of cell Ca. They may control it with high affinity (ATPases) or with lower affinity (a number of other systems). Endo(sarco)plasmic reticulum is responsible for the fine regulation of cytosolic Ca, whereas the mitochondrial transporting systems essentially control intramitochondrial Ca. The mitochondrial systems, however, also protects the cytosol against pathological Ca increases. The nucleus is now emerging as a prime actor in calcium signalling: important nuclear functions, beginning with the transcription of some genes, are controlled by calcium. The nuclear envelope contains a calcium pump and calcium channels modulated by IP3 and cADPr. Together with the nuclear pores, these systems of the envelope are likely to control the exchange of calcium between the cytoplasm and the nucleoplasm.

5. SUMMARY

The signalling function of calcium is controlled by proteins that bind it reversibly and specifically. Most of the soluble proteins belong to the EF-hand family. They decode calcium information by changing conformation twice, once upon complexing calcium and later upon interacting with target enzymes. Hundreds of these proteins are known, the most important of them being calmodulin. Other calcium controlling proteins are membrane-intrinsic (plasma membranes and organelles), and interact with calcium with high affinity (pumps) or with low affinity (sodium/calcium exchangers, channels, the electrophoretic uptake uniporter of mitochondria) pumps perform the fine tuning of cellular calcium, the plasma membrane sodium-calcium exchange is responsible for the lower affinity, bulk regulation of calcium. In heart and skeletal muscles, however, the calcium pump of sarcoplasmic reticulum is very abundant and thus also moves large amounts of calcium. The long-term, low affinity calcium regulation in situations of pathological increases of calcium entry, is performed by the mitochondrial uptake/release systems. A recent addition to the area of calcium signalling is the nucleus, which contains numerous calcium-modulated processes, e.g., the transcription of several genes. Nuclear calcium is controlled by the concerted operation of the nuclear pores and of several transporting systems in the envelope.

6. ACKNOWLEDGMENTS

The work described has been supported by the Swiss National Science Foundation (grant no. 31 30858.91).

7. REFERENCES

Babu, Y.S., Sack, J.S., Greenhough, T.J., Bugg, C.E., Means, A.R., & Cook, W.J. (1985). Three-dimensional structure of calmodulin. Nature, 315: 37–40.
Barton, G.J., Newman, R.H., Freemont, P.S., & Crumpton, M.J. (1991). Amino acid sequence analysis of the annexin super-gene family of proteins. Eur. J. Biochem., 198: 749–760.
Berridge, M.J. (1993). Inositol trisphosphate and calcium signalling. Nature, 361: 315–325.
Bewley, M.C., Bonstead, C.M., Walker, J.H., Waller, D.A., & Huber, R. (1993). Structure of chicken annexin Va at 2.25-Å resolution. Biochemistry, 32: 3923–3929.
Carafoli, E. (1979). The calcium cycle of mitochondria. FEBS Letts., 104: 1–5.
Carafoli, E. (1982). The transport of calcium across the inner membrane of mitochondria. In Membrane Transport of Calcium. Edited by Carafoli E. London: Academic Press, 109–139.
Carafoli, E. (1987). Intracellular calcium homeostasis Ann. Rev. Biochem., 56: 395–433.
Carafoli, E. (1991). Calcium pump of the plasma membrane. Physiol. Rev., 71: 129–153.
Carafoli, E. (1992). The Ca^{2+} pump of the plasma membrane. J. Biol. Chem., 267: 2115–2118.
Carafoli, E., Tiozzo, G., Lugli, F., Crovetti, F., & Kratzing, C. (1974). The release of calcium from heart mitochondria by sodium. J. Molec. Cell. Cardiol., 6: 361–371.
Concha, N.O., Head, J.F., Kaetzel, M.A., Dedman, J.R., & Seaton, B.A. (1993). Rat annexin V crystal structure: Ca^{2+} induced conformational changes. Science, 261: 1321–1324.
Cornwell, T.L., Prywantky, K.B., Wyatt, T.A., & Lincoln, T.M. (1991). Regulation of sarcoplasmic reticulum protein kinase phosphorylation by localized c-GMP-dependent protein kinase in vascular smooth muscle cells. Mol. Pharmacol., 40:923–931
Crompton, M., Sigel, E., Salzmann, M., & Carafoli, E. (1976). A kinetic study of energy-linked influx of Ca^{2+} into heart mitochondria. Eur. J. Biochem., 69: 429–434.
Crumpton, M.J., & Dedman, J.R. (1990). Protein terminology tangle. Nature, 345: 212.
Fabiato, A., & Fabiato, F. (1975). Contractions induced by a calcium triggered release of calcium from the sarcoplasmic reticulum of single skinned cardiac cells. J. Physiol., 249: 469–495.
Fleckenstein, A. (1973). Calcium antagonism in heart and smooth muscle. John Wiley, New York.
Furuichi, T., Yoshikawa, S., Miyawaki, A., Wada, K., Maeda, N., & Mikoshiba, K. (1989). Primary structure and functional expression of the inositol 1,4,5-triphosphate-binding protein P400. Nature, 342: 32–38.
Galione, A. (1993). Cyclic ADPribose, a new way to control calcium. Science, 259: 325–328
Herzberg, O., & James, M.N.G. (1985). Structure of the calcium regulatory muscle protein troponin-C at 2.8. A resolution. Nature, **313**: 653–659.
Hofmann, F., Anagli, J., Carafoli, E., & Vorherr T. (1994). Phosphorylation of the calmodulin binding domain of the plasma membrane Ca2+ pump by protein kinase C reduces its interaction with calmodulin and with the receptor site in the pump. J. Biol. Chem. (in press).

Hofmann, F., Biel, M., & Flockerzi, V. (1994). Molecular basis for Ca^{2+} channel diversity. Ann. Rev. Neuroscience, 17: 399–418.

Huber, R., Römisch, J., & Paques, E.P. (1990). The crystal and molecular structure of human annexin V, an anticoagulant protein that binds to calcium and membranes. EMBO J., 9: 3867–3874.

Huber, R., Schneider, M., Mayr, I., Römisch, J., & Paques, E.P. (1990). The calcium binding sites in human annexin V by crystal structure analysis at 2.0 Å resolution. FEBS Lett., 275: 15–21.

Ikura, M., Marius, Clore, G., Gronenborn, A.M., Zhu, G., Klee, C.B., & Bax, A. (1992). Solution structure of a calmodulin-target peptide complex by multidimensional NMR. Science, 256: 632–638.

James, P., Inui, M., Tada, M., Chiesi, M., & Carafoli, E. (1989). Nature and site of phospholamban regulation of the Ca^{2+} pump of sarcoplasmic reticulum. Nature, 342: 90–92.

Kofuii, P., Lederer, W.J., & Schulze, D.H. (1994). Mutually exclusive and cassette exons underlie alternatively spliced isoforms of the Na/Ca exchanger. J. Biol. Chem., 269: 5145–5149.

Kretsinger, R.H., & Nockolds, C.E. (1973). Carp muscle calcium-binding protein. J. Biol. Chem., 248: 3313–3326.

MacLennan, D.H. (1970). Purification and properties of an adenosine triphosphatase from sarcoplasmic reticulum. J. Biol. Chem., 245: 4508–4513.

MacLennan, D.H., Brandl, C.J., Korczak, B., & Green, N.M. (1985). Amino acid sequence of a Ca2+ + Mg2+ dependent ATPase from rabbit muscle sarcoplasmic reticulum deduced from its complementary DNA sequence. Nature, 316: 696–700.

Matsuoka, S., Nicoll, D.A., Reilly, R.F., Hilgemann, D.W., & Philipson, K.D. (1993). Initial localization of regulatory regions of the cardiac sarcolemmal Na^{+}-Ca^{2+} exchanger. Proc. Nat. Acad. Sci. (USA), 90: 3870–3874.

Meador, W.E., Means, A.R., & Quiocho, F.A. (1992). Target enzyme recognition by calmodulin: 2.4 Å structure of a calmodulin-peptide complex. Science, 257: 1251–1255.

Nicoll, D.A., Longoni, S., & Philipson, K.D. (1990). Molecular cloning and functional expression of the cardiac sarcolemmal Na^{+}-Ca^{2+} exchanger. Science, 250: 562–565.

Reiländer, H., Achilles, A., Friedel, U., Maul, G., Lottspeich, F., & Cook, N.J. (1992). Primary structure and functional expression of the Na/Ca, K-exchanger from bovine rod photoreceptors. EMBO J. , 11: 1689–1695.

Reuter, H. (1984). Ion channels in cardiac cell membranes. Ann. Rev. Physiol., 46: 473–484.

Reuter, H., Stevens, C.F., Tsien, R.W., & Yellen, G. (1982), Properties of single calcium channels in cardiac cell culture. Nature, 297: 501–504.

Richter, C. (1992). Mitochondrial calcium transport. In: New Comprehensive Biochemistry . Edited by Neuberger A, Van Deenen LLM., Elsevier: Amsterdam, 349–358.

Rosen, L.B., Ginty, D.D., & Greenberg, M.E. (1995), Calcium regulation of gene expression. Adv. Sec. Messengers and Phosphoprot. Res., 30: 225–253.

Santella, L. (1996). The cell nucleus: an eldorado to future calcium research? J. Membr. Biol, in press.

Shull, G.E., & Greeb, J. (1988). Molecular cloning of two isoforms of the plasma membrane Ca2+ transporting ATPase from rat brain. Structural and functional domains exhibit similarity to Na+, K+ and other cation transport ATPases. J. Biol. Chem., 263: 8646–8657.

Stauffer, T., Hilfiker, H., Carafoli, E., & Strehler, E.E. (1993). Quantative analysis of alternative splicing options of human plasma membrane calcium pump genes. J. Biol. Chem., 268: 25993–26003.

Streb, H., Irvine, R.F., Berridge, M.J., & Schulz, I. (1983). Release of Ca2+ from a non-mitochondrial intracellular store in pancreatic acinar cells by inositol-1,4,5-trisphosphate. Nature, 306: 66–69.

Szebenyi, D.M.E., Obendorf, S.K., & Moffat., K., (1981). Structure of vitamin D-dependent calcium binding protein from bovine intestine. Nature, 294: 327–332.

Tada, M., Kirchberger, M.A., & Katz, A.M. (1975). Phosphorylation of 22.000-Dalton component of the cardiac sarcoplasmic reticulum by adenosine 3":5"-monophosphate-dependent protein kinase. J. Biol. Chem., 250: 2640–2647.

Takeshima, H., Nishimura, S., Matsumoto, T., Ishida, H., Kangawa, K., Minamino, N., Matsuo, H., Ueda, M., Hanaoka, M., Hirose, T., & Numa, S. (1989). Primary structure and expression from complementary DNA of skeletal muscle ryanodine receptor. Nature, 339: 439–445.

Vaghy, P.L., Johnson, J.D., Matlib, M.A., Wang, T., & Schwarz, A. (1982). Selective inhibition of Na^{+}-induced Ca^{2+} release from heart mitochondria by diltiazem and certain other Ca^{2+} antagonist drugs. J. Biol. Chem., 257: 6000–6002.

Verma, A.K., Filoteo, A.G., Stanford, D.R., Wieben, E.D., Penniston, J.T., Strehler, E.E., Fischer, R., Heim, R., Vogel, G., Mathews, S., Strehler-Page, M.A., James, P., Vorherr, T., Krebs, J., & Carafoli, E. (1988). Complete primary structure of a human plasma membrane Ca2+ pump. J. Biol. Chem., 263: 14152–14159.

Weng, X., Luecke, H., Song, I.S., Kang, D.S., Kim, S.H., & Huber, R. (1993). Crystal structure of human annexin I at 2.5A resolution. Prot. Sci. 2: 448–458.

CALCIUM CONCENTRATION MICRODOMAINS

Rodolfo Llinás,[1] Mutsuyuki Sugimori,[1] and Robert Silver[2]

[1]Department of Physiology and Neurosciences
New York University Medical Center
550 First Avenue, New York, New York 10016
[2]Section and Department of Physiology
Cornell University
Ithaca, New York 14853-6401

1. INTRODUCTION

A question often posed in the field of neuroscience is that of the mechanism by which extracellular calcium releases transmitter from the presynaptic terminal (Katz & Miledi, 1965). Results from two decades ago, indicated that spatially restricted zones of transient high calcium concentration in the neuronal cytosol, could be produced as a result of the opening of voltage gated calcium channels in the plasmalemma (Llinás, 1977). This hypothesis together with the finding that the latency for calcium activation of transmitter release was in the range of 200 msec, suggests that calcium, at the presynaptic active zone, acts at a location very close to the release site and that the calcium channels themselves might be part of this active zone (Llinás, 1977).

Following presynaptic voltage clamp studies that described I_{Ca} and its relationship to transmitter release (Llinás et al., 1981a,b), a more formal approach was taken to understand the role of calcium in transmitter release. The regulation of cellular events by intracellular calcium had in the past been correlated with the effects of changing $[Ca^{2+}]o$, I_{Ca}, or with bulk cytosolic $[Ca^{2+}]$. Today, such correlations are known to be froth with difficulty. We now know, as predicted by quantitative models (Chad & Eckert, 1984; Simon & Llinás, 1985; Fogelson & Zucker, 1985), that calcium concentration changes occur in a punctate manner at those sites where voltage or ligand dependent calcium gates are present. It is also known that such intracellular calcium concentration profiles generate and disperse quite rapidly as the channels open and close, respectively. In addition, based on a set of studies conducted by Simon & Llinás (1985), it was proposed that given the diffusion coefficient of calcium, the magnitude of single channel calcium flow, as well as the calcium buffering ability of the cytoplasm, calcium concentration "microdomains" (i.e. sites in the cytoplasm that will show high calcium concentration profiles) would occur juxtaposed to an open calcium channel. These microdomains should be localized to the active zone, a region corresponding to the intra-membrane particles observed in

Calcium and Cellular Metabolism: Transport and Regulation, edited by Sotelo and Benech.
Plenum Press, New York, 1997

17

freeze fracture (Pumplin et al., 1981) and would be the trigger mechanism for transmitter release.

2. MODELING RESULTS

The results of these modeling studies not only provide a consistent description of many aspects of synaptic release, but also serves to highlight the two most significant features that define the importance of calcium concentration microdomains: their strategic location and the magnitude and speed of the calcium profile.

In this model (Simon & Llinás 1985) we provide a description of the three-dimensional diffusion of calcium entering through a single opened channel using three different numerical techniques: (1) treating the calcium as diffusing through a series of concentric hemispheric shells; (2) reducing the spherical diffusion equation:

$$dC(r,t)dt = D_{Ca} \times (2/r \times dC(r,t)/dr + d^2C(r,t)/dr^2)$$

from three spatial dimensions to one by substituting

$$U(r,t) = r \times C(r,t)$$

$$dU(r,t)/dt = D_{Ca} \times d^2U(r,t/dr^2)$$

where $C(r,t)$ is the calcium concentration as a function of time (t) and distance (r) from the channel pore and D_{Ca} the diffusion coefficient for calcium (6.0×10^{-6} cm^2/s); and (3) modeling the calcium as if it were being injected into one compartment of a three-dimensional Cartesian matrix.

The first two approaches utilized an implicit numerical iteration scheme. An explicit iteration was used for the three-dimensional diffusion matrix. To assess the robustness of the results the size of both the temporal and spatial steps were varied from 10^{-11} to 10^{-8} s and 1 to 25 Å, respectively.

3. MICRODOMAINS PRODUCED BY SINGLE CHANNEL INFLUX

A three-dimensional calcium isoconcentration surface generated during a channel opening is shown in Fig. 1A (Simon & Llinás, 1985). This structure is but one of a set of concentrational hemispheres that, like explosions, rapidly fill the aqueous space and come to a dynamic equilibrium for the duration of the calcium source flux. The calculated values of such steady state $[Ca^{2+}]_i$ are plotted as a function of distance from the channel pore for depolarizations of 35 mV (Fig. 1C). A cross section of this concentration profile is plotted in Fig. 1B ith the steady state calcium distribution in the presence of a saturable, mobile buffer (bars) superimposed on the same graph.

The transient and steady state $[Ca^{2+}]_i$ after the opening of the channel are plotted for a distance ("r") 500 Å from the pore (Fig. 1D). Over the first few hundred angstroms from the channel pore the concentration is close to steady state within a microsecond of channel opening at close to 400 mM. Following channel closing, if one assumes that the calcium concentration decrease is due to diffusion, the peaks of $[Ca^{2+}]_i$ near the channel disappear in less than 1m[m]s (Fig. 1E).

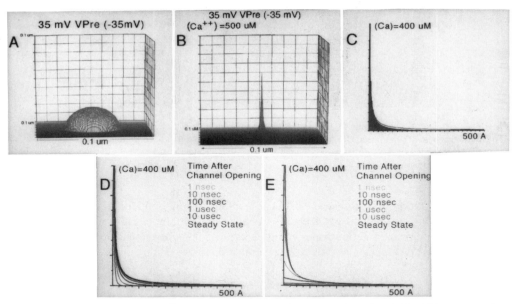

Figure 1. Steady state cytosolic $[Ca^{2+}]_i$ during the open time of a calcium channel. (A) 10 mM isoconcentration surface for $[Ca^{2+}]_i$. (B) The $[Ca^{2+}]_i$ as a function of distance from the pore in the absence (orange) and presence (blue) of a mobile buffer. (C) The $[Ca^{2+}]_i$ in the two-dimensional surface immediately under the membrane is plotted for depolarizations of 35 mV from a resting potential of −70 mV. (D) Transient and steady state $[Ca^{2+}]_i$ after calcium channel opening, plot for the first 500 Å$[Ca^{2+}]_i$, within a few hundred Angstroms of the channel, reaches a steady state within microseconds. (E) $[Ca^{2+}]_i$ after channel closing calculated with the assumption that the $[Ca^{2+}]_i$ decreased only as a result of diffusion for the first 500 Å. Time after channel opening and closing at the right of D and E (modified from Simon and Llinás, 1985).

4. EXPERIMENTAL DEMONSTRATION

The methodology available for the identification of such microdomains is still fairly limited. Most of the fluorescent calcium sensitive dyes do not have the necessary properties to localize in space these minute calcium profiles. Having worked with aequorin in the past we felt that the use of a rather insensitive aequorin capable of responding only to calcium concentration in the order of 300 to 400 μM would be the right tool to use. Such aequorin was provided by Drs. Shimomura and Kishi in the form of a synthetic protein which they called *n*-aequorin-J. To our delight, once the terminal was fully loaded with this protein (Fig. 2A) and the presynaptic fiber was repetitively stimulated, a well-defined, stable set of quantum emission domains (QEDs) appeared as short-lived bright spots (Fig. 2B) (Llinás et al., 1992). Superposition of the fluorescence images of the terminal digit (Fig. 2A) and the QEDs (Fig. 2C) revealed that the distribution of QEDs coincided with the probable distribution of active zones at the presynaptic terminal.

Each QED fell within a contiguous rectilinear juxtaposition of approximately 16 pixels (0.25 mm by 0.25 mm per pixel). The size of the microdomains, determined by measuring more than 15,000 events, fluctuated from 0.25 to 0.6 mm^2 (mean 0.313 mm^2, range, ~0.25 mm^2 to ~0.375 mm^2).

These microdomain sites occupied, on an average, 8.4% of the presynaptic-postsynaptic membrane contact area, within the range of the 5 to 10% determined by ultrastruc-

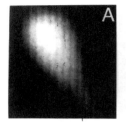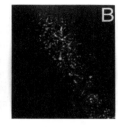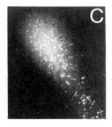

Figure 2. (A) Fluorescence image of a presynaptic digit injected with a fluorescent preparation of *n*-aequorin-J. (B) Photons image of superposed QEDs. Each dot, a microdomain represents the accumulated light emission from *n*-aequorin-J elicited by Ca^{2+} entry during tetanic stimulation. (C) Superposition of the images in (B) and (C). (modified from Llinás *et al.*, 1992).

tural studies (Pumplin & Reese, 1978). The number of microdomains in a 70 by 40 mm contact area was estimated by our results to be about 4500, quite close to the 4400 calculated for the number of active zones (range 3580 to 5400) from measurements and analysis of transmission electron micrographs (Pumplin & Reese, 1978).

5. CALCIUM MICRODOMAIN TIME COURSE

As transmitter release itself is triggered by a calcium binding step involving a low-affinity Ca^{2+} site at the presynaptic active zone, we expected calcium microdomains to have a very short time course.

The time course of these high concentration sites was analyzed in the squid giant synapse using rapid video images (4000 video frames sec^{-1}) following presynaptic injection of photoprotein *n*-aequorin-J. Microdomains evoked by presynaptic spike activation had an average duration of approximately 800 ms (Llinás et al., 1995). Spontaneous quantum emission domains (QEDs) observed during periods lacking stimulation occurred at about the same locations as the evoked microdomains, but of a lower emission intensity. The spontaneous QEDs observed prior to stimulation were imaged, as was the evoked release, over an area corresponding to ~20% of the total active zone of the presynaptic terminal studied. They occurred at a very low rate (10–15 sec^{-1}) and were generally located at the same sites as the evoked microdomains.

These results indicate that the time course of the calcium concentration profiles responsible for transmitter release is of extremely short duration and compares closely with that of calcium current flow during a presynaptic action potential (Llinás et al., 1982). This further demonstrates that, as theorized in the past (Chad & Eckert, 1984; Simon & Llinás, 1985; Fogelson & Zucker, 1985), intracellular calcium concentration at the active zone remains high only for the duration of transmembrane calcium flow.

An additional observation from that work was the realization that spontaneous and evoked light emissions probably correspond to spontaneous (Silver et al., 1994; Sugimori et al., 1994) and voltage-evoked calcium channel openings, respectively. The low frequency of such photon emissions indicates, however, that only a very small percentage of the total number of calcium entry events are being observed. In fact, for spontaneous release given calculated levels of 35,000 quanta per sec for the total active zone (Silver et al., 1994; Sugimori et al., 1994), the area observed (20%) should generate approximately 7000 QED per sec rather than the 10–15 observed assuming that one activated zone releases one QED. This number is based on several hypothesis: that (i) every spontaneous

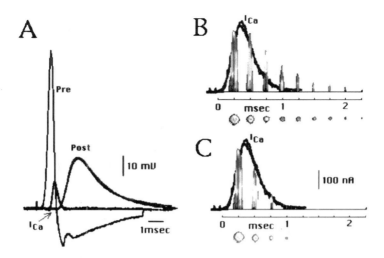

Figure 3. Time course for evoked calcium concentration microdomains. (A) Illustration of the time course for a presynaptic action potential (Pre) that simulates the voltage profile of a presynaptic spike recorded from the same fiber, before block of sodium and potassium channels (from Llinás *et al.*, 1982); the calcium current (I_{Ca}) generated by the transient depolarization; and the postsynaptic response (Post) triggered by I_{Ca}. The presynaptic current (I_{Ca} - arrow), basically a tail current with an overall duration of ~800 msec, and the postsynaptic potential obtained in previous experiments (Llinás *et al.*, 1982 are shown). (B and C) Superposition of the time course for the calcium microdomain (amplitude on a relative scale) and the time course for the calcium current. (B) Amplitude of a raw light measurements taken every 250 msec. (C) Amplitudes after correction for afterglow. The area of the circles under each time point relates to the amplitude of the light emission during that time bin. (Modified from Sugimori *et al.*, 1994).

event is triggered by calcium entry at the active zone (which will overestimate the expected QEDs because some spontaneous transmitter release is calcium independent), and (ii) each calcium entry event activates aequorin (which is also an overestimate because such reactions, being dependent on concentration and time, are far from fully efficient).

The QED frequency for evoked release occurred at a rate of 250–500 per sec, with superimposed QED occurring at a rate of 3 to 5 per sec. These values were 20 to 30 times greater than the QED frequency for spontaneous activity. The QED frequency for the relative increase in evoked release was at about the level expected, assuming a release of 5,000 to 10,000 quanta with each action potential (Sugimori et al., 1994). With one second of tetanic stimulation at 100 Hz, the quantal release should be around 0.5 to 1×10^6.

These experimental results indicate detection levels three orders of magnitude smaller than expected from such calculations. Two possible explanations for this very low yield are that QEDs are generated only when calcium channel openings occur with a particular spatial distribution (e.g., clustering), i.e., the so-called "neighbor effect" (Simon & Llinás, 1985); or that our detection system sees only a small percentage of the photons emitted (Silver et al., 1994). Measurements suggest that only 0.22% of the light emitted is captured by the present system (Sugimori et al., 1994), which supports the view that the low observed QED yield is due, at least in part, to technical limitations in detection (Silver et al., 1994).

In order to compare the time course for evoked QED with that of the presynaptic calcium current triggered by a simulated action potential, these events were superimposed on the same time base (Fig. 3). The time course for photon emission of evoked QED superimposes adequately on the time course of a microdomain evoked during the activation of similar presynaptic spikes (Fig. 3B). Note, however, that the light emission seems to outlast that of the transmembrane current. This is to be expected because high-level light emission leaves an afterglow in the photosensors (Silver et al., 1994). In fact, as shown in Figure 3 panel C, the time course for the microdomain corrected for this distortion (Sugimori et al., 1994) matches quite closely the time course of the macroscopic I_{ca}.

6. DISCUSSION

The results of the aequorin studies show us the true time course for calcium microdomains at a chemical presynaptic terminal. These findings are also in agreement with descriptions of the time course for spike-evoked calcium currents (Llinás et al., 1982) and with results from model studies (Llinás et al., 1981; Llinás et al., 1982) concerning the distribution and time course of transient calcium concentration profiles. Our calculations of the calcium-concentration profile indicated that a steady-state $[Ca^{2+}]_i$ level of approximately 300 μM was attained at 500 Å from the channel pore in 1 μs and, following channel opening, and that the concentration profile decays to the control level in a similar period, for the same distance (Simon and Llinás, 1985). Given the sensitivity level of *n*-aequorin-J, our results are, indeed, consistent with the model.

The time course of microdomains at the squid giant synapse differs from that observed in other secretory systems such as the chromaffin cell (Monck et al., 1994), where the transient develops over 10 to 100 ms and where the calcium buffering properties may be quite different. However in this case, calcium buffering by the fluorescent calcium indicators used may have contributed to the longer time courses reported. The squid calcium microdomain dynamics are, on the other hand, consistent with calcium concentration fluctuations monitered in individual frog-muscle sarcomeres following the onset of action potentials (Monck et al., 1994).

The overall duration of the microdomain is in agreement with the increased probability of release produced by a presynaptic spike. The calcium current produced by a presynaptic action potential has a peak at 0.3 ms and an overall duration of about 0.8 ms (Sugimori et al., 1994), very much in agreement with measurements observed with the present technique.

Viewed from another perspective, the demonstrated time course for calcium concentration microdomains matches closely that of the release process (Llinás et al., 1982) and the average open time duration of the P-type calcium channels (Usowicz et al., 1992) that are responsible for the presynaptic calcium current in this terminal (Llinás et al., 1989).

The number of calcium ions that are detected in an evoked microdomain may be estimated as follows. Consider that a presynaptic action potential generates an I_{Ca} of about 300 nA (Llinás et al., 1982), a current of 0.5 pA/channel, and that a single channel allows the flow of about 150 to 200 calcium ions (Pumplin et al, 1981; Llinás et al 1982; Stanley, 1993) then roughly 6×10^5 channels are opened to release 5000 to 10,000 vesicles or approximately 15,000 calcium ions per vesicle (Llinás et al., 1992). This is in contrast to the estimated minimum of 200 ions/vesicle (one calcium channel) for the ciliary ganglion (Stanley, 1993). Given that the number of active zones in a presynaptic terminal is about 5000 to 10,000 (Pumplin and Reese, 1978), the number of channels open per action potential for each active zone is about 100. If, as mentioned, single channels allow the flow of 150 to 200 calcium ions, then using *n*-aequorin-J and the present imaging technique an evoked microdomain may represent an influx of calcium of about 15,000 ions per action potential (Llinás et al., 1995). This is equivalent to the release of a single vesicle at an active zone upon elevation of the local calcium concentration to 3.8×10^{-4} M, in agreement with previous measurements (Llinás et al., 1982; Simon and Llinás, 1985). Indeed, the $[Ca^{2+}]$ we measured coincides with the local $[Ca^{2+}]$ requirements for the newly discovered family of proteins responsible for docking of presynaptic secretory vesicles with the plasma membrane (Perin et al., 1990).

7. REFERENCES

Chad, J.E. & Eckert, R. (1984).Calcium domains associated with individual channels may account for anomalous voltage relation of Ca-dependent response. Biophys. J. 45:993–1000.

Fogelson, A.L. & Zucker, R.S. (1985). Presynaptic calcium diffusion from various arrays of single channels. Implication for transmitter release and synaptic facilitation. Biophys. J. Vol. 48: 1003–1017.

Katz, B. & Miledi, R. (1965). The measurement of synaptic delay and the time course of acetylcholine release at the neuromuscular junction. Proc. Royal Soc. Lond. (Biol.) Vol. 161:483–495.

Llinás, R. (1977). Calcium and transmitter release in squid synapse. In: Society for Neuroscience Symposia, eds. W.M. Cowan and J.A. Ferendelfi, Bethesda: Society for Neuroscience, 2:139–160.

Llinás, R., Steinberg, I.Z. & Walton, K. (1981a). Presynaptic calcium currents in squid giant synapse. Biophys. J. 33:289–322.

Llinás, R., Steinberg, I.Z. & Walton, K. (1981b). Relationship between presynaptic calcium current and postsynaptic potential in squid giant synapse. Biophys. J. 33:323–352.

Llinás, R., Sugimori, M., Lin, J-W., Leopold, P. & Brady, S. (1989). ATP-dependent directional movement of rat synaptic vesicles injected into the presynaptic terminal of squid giant synapse. Proc. Natl. Acad. Sci. USA, 86:5656–5660.

Llinás, R., Sugimori, M. & Silver, R.B. (1992). Microdomains of high calcium concentration in a presynaptic terminal. Science, 256:677–679.

Llinás, R., Sugimori, M. & Silver, R.B. (1995). The concept of calcium concentration microdomains in synaptic transmission. Neuropharmacology, 34:1443–1451.

Llinás, R., Sugimori, M. & Silver, R.B. (1995). Time resolved calcium microdomains and synaptic transmission. J. Physiology (Paris) 89:77–81.

Llinás , R., Sugimori, M. & Simon, S.M. (1982). Transmission by presynaptic spike-like depolarization in the squid giant synapse. Proc. Natl. Acad. Sci. USA 79:2415–2419.

Monck, J.R., Robinson, I.M., Escobar, A.L., Vergara, J.L. & Fernandez, M. (1994). Pulsed laser imaging of rapid Ca^{2+} gradients in excitable cells. Biophys. J. 67:505–514.

Perin, M.S., Fried, V.A., Mignery, G., Jahn, R. & Südhof, T. C. (1990). Phospholipid binding by a synaptic vesicle protein homologous to the regulatory region of protein kinase C. Nature 345:260–263.

Pumplin, D.W. & Reese, T.W. (1978). Membrane ultrastructure of the giant synapse of the squid *Loligo pealei*. Neuroscience 3:685–696.

Pumplin, D.W., Reese, T.W. & Llinás, R. (1981). Are the presynaptic membrane particles the calcium channels? Proc. Natl. Acad. Sci. (USA) 78:7210–7218.

Silver, R.B., Sugimori, M., Lang, E.J. & Llinás, R. (1994). Time resolved imaging of Ca^{2+}-dependent aequorin luminescence of microdomains and QEDs in synaptic terminals. Biol. Bull. 187:293–299.

Simon, S. & Llinás, R. (1985). Compartmentalization of the submembrane calcium activity during calcium influx and its significance in transmitter release. Biophys. J. 48:485–498,.

Stanley, E.F. (1993). Single calcium channels and acetylcholine release at a presynaptic nerve terminal. Neuron 11:1007–1011.

Sugimori, M., Lang, E.J., Silver, R.B. & Llinás, R. (1994). High-resolution measurement of the time course of calcium-concentration microdomains at squid presynaptic terminals. Biol. Bull. 187:300–303.

Usowicz,M.M., Sugimori, M., Cherksey, B. & Llinás, R. (1992). P-type calcium channels in the somata and dendrites of adult cerebellar Purkinje cells. Neuron 9:1185–1199.

3

RAPID CONFOCAL MEASUREMENTS OF Ca^{2+} SPARKS IN RAT VENTRICULAR MYOCYTES

Lars Cleeman, Wang Wei, and Martin Morad

Georgetown University Medical Center
3900 Reservoir Road NW
Washington, DC 20007

1. INTRODUCTION

1.1 Calcium Release in Mammalian Cardiomyocytes

The initiation of contraction in mammalian myocardial cells begins with the entry of Ca^{2+} through sarcolemmal Ca^{2+} channels (DHP-receptors) and triggering of Ca^{2+} release from the sarcoplasmic reticulum (SR) via Ca^{2+} release channels (ryanodine receptors; Näbauer et al., 1989). At the level of optical resolution (0.2–0.5 μm) the functional unit may be perceived as one sarcomere (1.6 to 2.2 μm z-line to z-line) of a single myofibril (diameter $\cong$ 1 μm). This "sarcomeric unit" receives its Ca^{2+} from a collar of SR which forms dyadic junction with transverse tubules (t-tubules—invaginations of the surface membrane). Below the level of optical resolution, electron microscopy shows that each dyadic junction (0.1 μm diameter, about 30 nm wide) contains a number of DHP receptors in the t-tubular membrane and a larger number of ryanodine receptors opposing them in the SR membrane (Jorgensen et al., 1993; Carl et al., 1995). While there appear to be no mechanical contact between the two types of receptors in cardiac muscle (Sun et al., 1995), it is clear that their communication via rapid local rises in [Ca^{2+}]$_i$ (Sham, Cleemann and Morad, 1995; Adachi-Akahane, Cleemann and Morad, 1996) and activation of Ca^{2+}-induced Ca^{2+} release mechanism (CIRC; Fabiato, 1983, 1985) is of fundamental importance for the control of the cardiac contraction. Yet the details of this control process remains elusive. What is missing, in part, seems to be a clearer understanding of the way the DHP and ryanodine receptors interact in their normal environment. For instance, it is not clear whether the basic Ca^{2+} signaling unit (Ca^{2+} μ-domain) is a) a single ryanodine receptor responding to influx of Ca^{2+} via a single DHP receptor, b) a cluster of ryanodine receptors controlled by one DHP receptor, c) an entire dyadic junction with several DHP and Ryanodine receptors, or even d) an aborted Ca^{2+} wave spreading over several sarcomeric units.

Calcium and Cellular Metabolism: Transport and Regulation, edited by Sotelo and Benech.
Plenum Press, New York, 1997

2. MICRO-DOMAINS OF CALCIUM

Two strategies have been used to define the properties of Ca^{2+} μ-domains. One is to view them directly as "Ca^{2+} sparks" using Ca^{2+}-indicator dyes and confocal fluorescence microscopy. This method becomes less direct when it is considered that sensitivity and speed of the technique applied thus far are often marginal, as the processes in question may well take place over distances below the limits of optical resolution with visible or near UV light. Another approach, undertaken primarily by our laboratory, is to assess the size of the Ca^{2+} μ-domain by addition of Ca^{2+} buffers to the cytosol thereby greatly limiting the diffusion distance of Ca^{2+} (Stern, 1992). This approach, however, suffers from the criticism that the Ca^{2+} buffers may not gain access to the spaces which define the μ-domains, and that their kinetics may be too complicated to allow direct interpretation of the results (Smith, Wagner and Keizer, 1996).

The Ca^{2+} sparks are local, non-propagated, Ca^{2+} current-dependent, and ryanodine-sensitive Ca^{2+} releases which occur spontaneously in resting cardiac cells and have been detected with confocal microscopy either in the line scan mode (500–1000 Hz) at low frame rates where they appear in only a single frame (Niggli and Lipp, 1992; Cheng Lederer and Cannell 1993). It has been proposed that intracellular Ca^{2+} transients produced by action potentials or voltage clamp depolarizations result from summation of sparks (Lopez-Lopez, Shacklock, Balke and Wier, 1994; Cannell, Cheng and Lederer, 1995; Santana, Cheng, Gomez, Cannell and Lederer, 1996). Sparks measured at rest have been estimated to represents the release of about $2*10^{-19}$ moles of Ca^{2+} and decays with a time constant of 20–25 ms (Cheng *et al.*, 1993). Conceivably such a Ca^{2+} spark could result from a 10 ms opening of a single Ca^{2+}-release channel. It has also been observed, however, that the magnitude of individual sparks decreases as the level of voltage clamp depolarizations are changed from −30 to +50 mV (Lopez-Lopez *et al.*, 1994) and that detectable Ca^{2+} sparks are absent when cellular Ca^{2+} releases are triggered with caged Ca^{2+} (Lipp and Niggli 1996), suggesting that Ca^{2+} sparks may not represent the fundamental unitary events in excitation-contraction coupling.

In an alternative approach to subcellular Ca^{2+} signaling, millimolar concentrations of Ca^{2+} buffer (2 mM Fura-2 or 2 mM Fura-2 together with 14 mM EGTA or 10 mM BAPTA) have been used to dialyze ventricular myocytes in order to limit the effective diffusion distance of Ca^{2+} to 10 to 50 nm. Under these conditions activation of myofilaments, efflux of Ca^{2+} on the Na^+-Ca^{2+} exchanger, and significant rise in cytoplasmic Ca^{2+} concentration are blocked or strongly suppressed, but the uptake of Ca^{2+} by the SR, its release via the ryanodine receptor, and alteration of the kinetics of the Ca^{2+} current in response to the local rise of Ca^{2+} remain intact (Adachi-Akahane *et al.*, 1996). Addition of Ca^{2+} buffer in millimolar concentrations, in fact, makes it possible to estimate the amount of released Ca^{2+} and show that released Ca^{2+} remains constant at about 140 μM (in the dye-accessible intracellular space). This is about 18 times larger than the Ca^{2+} influx through the L-type channel. Conversely the L-type channel is responsive to the released Ca^{2+} as indicated by a 3 to 4 fold acceleration of its rate of inactivation. These findings support the idea that DHP- and ryanodine-receptors, in the absence of a direct mechanical connection, may be functionally coupled by two-way exchange of Ca^{2+} signals via μ-domains of Ca^{2+}. The major limitation of this approach is that the estimation of diffusion distance based on buffering capacity is hardly sufficiently accurate to determine the exact numbers of DHP- and ryanodine-receptors in a μ-domain, nor is it certain that the buffered cells retain the type of excitation-contraction coupling which give rise to Ca^{2+} sparks.

In the present study we have attempted to increase the spatial and temporal resolution of Ca^{2+} measurement in cardiac cells by using a rapidly scanning confocal microscope in conjunction with internal dialysis of millimoles of Ca^{2+} dyes and buffers. Our results indicate that this approach is largely successful in recording the detailed properties of Ca^{2+} sparks. We find that Ca^{2+} sparks are small and well defined (about 1 μm in diameter) and originate from regions near the z-lines. The release lasts at most 8 ms and may prove to be even shorter when higher frame rates are tested (>120 Hz). The decay or dispersion of Ca^{2+} sparks is also largely complete in about 8 ms. the combination of the two approaches provide a powerful technique for defining and quantifying the fundamental steps in cardiac excitation contraction coupling.

3. DYNAMIC MEASUREMENTS OF Ca²⁺ SPARKS WITH RAPID CONFOCAL MICROSCOPY

Ca^{2+} sparks were measured with a rapidly scanning (line frequency 17 or 35 KHz, 100 ns/pixel), acousto-optically steered, confocal microscope (Noran, Odyssey XL). While previous investigations with confocal microscopy have measured the Ca^{2+} sparks along a single line (line scan mode, line frequency 500 Hz to 1 KHz, 1000 to 2000 ns/pixel), we used a raster pattern and collected 30 or 120 frames per second. It was thought that the two dimensional fluorescence maps produced in this way would help to define the detailed properties of the Ca^{2+} sparks and locate them in the context of the cellular architecture.

The confocal equipment was mounted on an inverted microscope (Zeiss, Axiovert 135 TV) with a water immersion objective (Zeiss, C-apochromat, 40x, NA: 1.2w) designed for measurements as far as 100 μm from the bottom of the chamber. The excitation beam was the 488 nm line from an Argon ion laser (Omnichrome), the fluorescent Ca^{2+} indicator dye was Fluo-3 (Molecular Probes), the confocal slit was set at 25 μm (0.4 μm in the focal plane), and the fluorescent light was measured with a high efficiency photo multiplier tube (Hamamatsu). The data were collected by a computer (Silicon Graphics, Indy, UNIX operating system) and stored temporarily in 100 Mbytes of random access memory before it was transferred to hard disk and finally to rewritable optical disk. In some experiments a patch clamp amplifier (DAGAN) was used in the whole cell configuration together with a second computer (IBM compatible, pCLAMP software) to establish internal dialysis and control the resting potential (Hamill et al, 1981).

4. CALCIUM WAVES IN CELLS STAINED WITH FLUO-3 AM

In initial experiments rat ventricular myocytes cells were loaded with Ca^{2+} indicator dye (without internal dialysis) by incubating with 10 μM Fluo-3-AM for 30 to 60 min. Such cells showed Ca^{2+} transients of three distinct types: a) discrete Ca^{2+} sparks, b) aborted local waves, and c) waves spreading throughout the cells. The Ca^{2+} sparks had typical dimensions of 1 to 3 μm, had roughly circular symmetry, came on rapidly and faded away more slowly (in less than 100 ms), and were seen most clearly in heavily stained cell and with a relatively high intensity of the scanning beam (which might bleach the dye with a time constant as low as 10 to 5 sec). Most sparks appeared to occur randomly, but in some cases it was possible to identify hyperactive sites where sparks appeared to repeat at a frequency of 1 to 4 Hz. An aborted wave typically originated from a point but died out after spreading a few microns, probably because SR loading was marginal for wave propagation. Wave propagation throughout a cell often occurred at regular intervals from an active focus near the edge of the cell. Waves propagating from competing foci frequently collided and extinguished each other thereby demonstrating refractoriness behind the wave front. This type of block might give rise to re-entry patters which might be regular (Fig. 1) or chaotic without any apparent pattern. The counter-clockwise spiral wave in Fig. 1 was stable for a number of cycles. It is seen very clearly in this unusually wide cell. The rapid rise at the wave front was not seen in the nuclear regions, which, though higher in resting fluorescence, only displayed attenuated and delayed Ca^{2+} transients (Cf. O'Malley, 1994). The overall pattern indicates that the wave fronts were not significantly retarded by the nuclear regions, most likely because they occupy only a small fraction of the total thickness of the cell. On the other hand, it may be noticed that spiral pattern does not have perfect circular symmetry but is stretched along the axis of the cell thereby suggesting, that the Ca^{2+} waves spread in the longitudinal direction about twice as fast (80 µm/sec) as in the transverse direction (40 µm/sec).

During continuous measurements, the cells often progressed through the three stages outlined above and ended in irreversible contraction. This suggest that extensive exposure to light not only bleached the dye, but also had toxic effects resulting in Ca^{2+} overload and cell death.

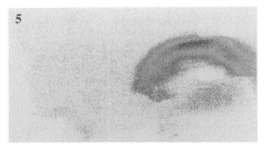

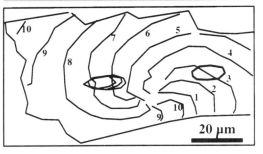

Figure 1. Spiral wave in a rat ventricular myocyte stained with Fluo-3 AM. The lower panel shows the outline of the cell, two nuclear regions and 10 wave fronts recorded at 133 ms intervals. The longitudinal axis of the cell is in the left-right direction as indicated by the sarcomere pattern, the contractions and the orientation of the nuclear regions. The upper panel shows the fluorescence images corresponding to wave fronts #5 and #8.

5. LOADING OF CELLS WITH FLUO-3 VIA THE PATCH PIPET

The following experiments were performed under voltage clamp control and Fluo-3 penta-potassium salt was introduced through the patch pipette. The time course of loading is illustrated in Fig. 2. In this experiment the time constant for equilibration of the cell by diffusion of the dye through the tip of the pipette was $\tau = 150$ s. In comparison, the longitudinal diffusion constant for the dye was estimated as $D = 0.25 \times 10^{-6}$ cm²/s which, in a cell with a length, L =150 μm, gives a typical time constant of $(L/\Pi)^2/D = 90$ s. In most cells the diffusion through the tip was even slower ($\tau = 300$ to 600 s). This mean that the rate limiting process in the equilibration process is diffusion through the tip of the pipette—not diffusion within the cell—and it calls for patch pipettes with relatively large tip and low resistance.

The distribution of Fluo-3 in the resting cell is illustrated in Fig 3. The outline of the cell is well defined and the average fluorescence intensity is only slightly larger near the center than at the edges suggesting a high degree of, but not perfect, confocality. Nuclear regions typically have lines with higher-than-average fluorescence intensity extending in the longitudinal direction. The most noticeable feature, however, is the sarcomere pattern which is often seen clearly, even though it only produces intensity variations of less than

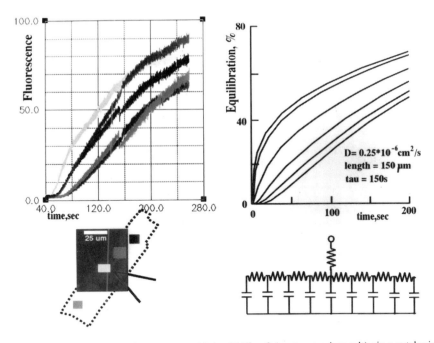

Figure 2. Loading of a rat ventricular myocyte with 1 mM Fluo-3 (pentapotassium salt) via a patch pipette. The lower left panels show the outline of the cell, the position of the patch pipette, and different regions of interest labeled by gray tones. The upper left panel shows fluorescence intensity in the different regions of interest as function of time following "break-in" at 50 sec. The right panels show results of a computer simulation of the diffusion processes governed by the equivalent diagram at the bottom. $\tau = 150$ sec is the time constant for equilibration into the cell through the tip of the pipette. The equilibration along the axis of the cell is determined by the length of the cell (150 μm) and the diffusion constant of the dye ($0.25 * 10^{-6}$ cm²/s). The upper- and lowermost curves correspond, respectively, to the exact tip of the pipette and very end of the cell (i.e. locations not subject to measurement).

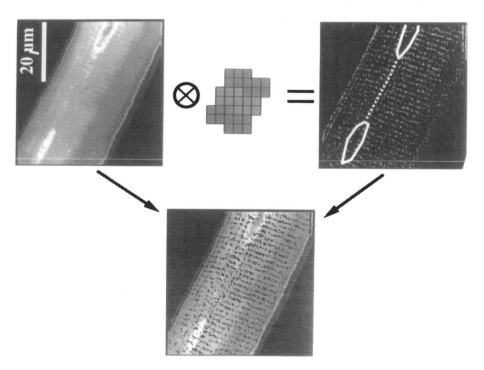

Figure 3. Fluorescence intensity in a resting cell. The upper left panel shows the fluorescence intensity averaged from 96 frames. This image was convoluted with the shown kernel where the central 11 pixels had a weight of 1 and the pixels on either side a weight of -0.5. The resulting image is shown in the upper right corner with the contrast enhanced to produce a black and white picture. The lower panel shows the two images overlaid.

10 %. The brighter parts of the fluorescence pattern corresponds to the z-lines as it has been found to coincide with extracellular staining of t-tubules (Shacklock *et al.*, 1995). It is an advantage, however, to be able to identify the sarcomere structure with Fluo-3 alone. For this purpose it was frequently useful to enhance the sarcomere pattern with an edge detector tuned to transverse ridges in the fluorescence intensity. This was done by convoluting the fluorescence images with kernels of the type shown in Fig. 3. This may pick out the sarcomere pattern in cells where the casual observer only sees random noise. Also, it can often identify locations where adjacent myofibrils appear to be out of register, as is the case here the case on a line connection the nuclear regions.

6. RANDOM LOCAL RELEASES OF Ca^{2+} (SPARKS)

Myocytes voltage clamped at a fixed resting potential between −60 and −100 mV had a low frequency of randomly occurring Ca^{2+} sparks. The top panel of Fig. 4 shows at least Ca^{2+} sparks in a single frame (#29) recorded from a cell held at −90 mV. Regions of interest were defined centered around a single prominent spark (arrows in frame #29) and their fluorescence intensity was followed in a total of 50 frames recorded at 30 Hz for a period of one and a half second (lower panel).

Ca^{2+} sparks generally appeared in one frame without any detectable warning in the previous frame, but had some carry over to the following 1 or 2 frames. This suggests that

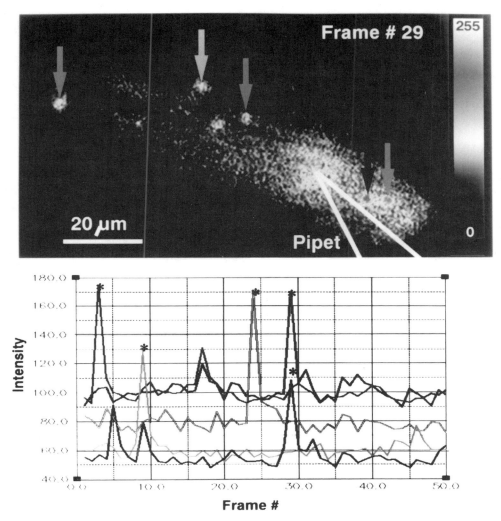

Figure 4. Ca^{2+} sparks in a cell held at -90 mV. The top panel shows the fluorescence intensity low-pass filtered with a 3 by 3 kernel. The higher intensity in the vicinity of the tip indicates that equilibration is incomplete. Locations of well defined Ca^{2+} sparks are marked with arrows and served to define regions of interest . The fluorescence intensity in the regions are shown in the lower panel for 50 consecutive frames in the lower panel. The dialyzing solution contained 1 mM Fluo-3 and 1 mM EGTA, and full frames (640×480 pixels) were recorded at 30 Hz.

the release occurs in less than 30 ms but that the reuptake of released Ca^{2+}, or its dispersion by diffusion, takes significantly longer time. The prominent sparks (*), which served to define the regions of interest, typically produced a 1.95 fold increase of fluorescence in a 3.8 μm^2 area. Earlier and later parts of the intensity traces shows a number of Ca^{2+} spikes of smaller amplitude. Inspection of the relevant frames (frames #5, #9, #17) showed that these transients, in general, are not smaller releases from the same locations but releases removed 1 to 2 μm and straddling the boundary of the region of interest.

The frequency of the Ca^{2+} sparks was increased when the holding potential was changed from −90 to −60 mV. This increase in the frequency of Ca^{2+} sparks with partial depolarization was a consistent finding observed in 6 cells where the holding potential was varied in the range of voltages from -100 to -60 mV, a range identified to be negative to activation of whole cell L-type Ca^{2+} channel.

To improve the temporal resolution we recorded "quarter frames" (320 by 240 pixels covering an area of 64 by 48 µm) at a rate of 120 Hz. Furthermore, we used a dialyzing solution which, in addition to 1 M Fluo-3, contained 5 mM EGTA in hope of further improving resolution of localized rise in $[Ca^{2+}]_i$. The idea is that Ca^{2+} will bind first to Fluo-3 and then be transferred to EGTA. This would accelerate the decay of the Ca^{2+} sparks, so that it would be governed mainly by the off-rate of Fluo-3 (about 200/s). Theoretically this would allow only 5 ms $(=1/(200/s))$ for the Ca-Fluo-3 complex to spread by diffusion over a distance of about $\sqrt{D} \times t = \sqrt{0.25} \times 10^{-6} \times 5 \times 10^{-3}$ cm $= 0.35$ µm. This is near the limit of optical resolution.

Figure 5 shows 6 consecutive partial frames recorded using this approach. With 8 ms between frames it is interesting to notice that well defined Ca^{2+} sparks appear in different spots in frames 12 and 14 without any hints in the previous frames. The bright spots representing these sparks have a diameter of about 1 µm and their centers can probably be defined within a quarter thereof. The following frames, on the other hand show the remnants of the decaying sparks. It is an intriguing observation that these remnants often appear as brightness surrounding a central region where the intensity is already back to normal. This is the case, not only for the strong sparks in panels 12 to 16, but also for the fainter spark appearing in the lower left quadrant of frames 10 and 11. Possibly the Ca^{2+} sparks either are produced by Ca^{2+} release which has a tendency to spread to neighboring release sites, or the region of SR from which the release occurred is immediately active in re-sequestering of Ca^{2+} from its near surroundings. The rapid decay of the central region of a Ca^{2+} spark, from on frame to the next, indicate that the release is terminated in less than 8 ms.

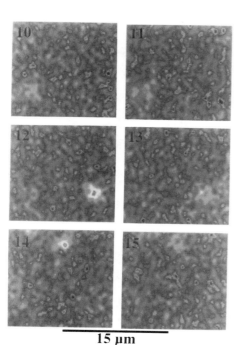

15 µm

Figure 5. Six consecutive frames recorded at 8 ms intervals in a rat ventricular myocyte dialyzed with 1 mM Fluo-3 and 5 mM EGTA. By chance and selection each row shows a single Ca^{2+} spark in two stages of development.

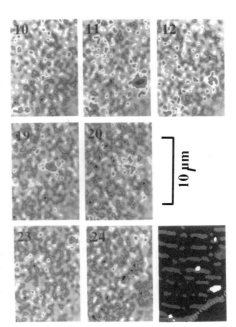

Figure 6. Ca^{2+} sparks and their relation to the sarcomere pattern in a rat ventricular cell dialyzed with 1 mM Fluo-3 and 10 mM EGTA. Each row shows 3 or 2 frames recorded consecutively at 8 ms intervals. The lower row in addition shows the sarcomere pattern (enhanced as illustrated in Fig. 3), the locations of the release sites (white), and the edge of the cell (dashed line).

The high spatial resolution of these measurements suggest that it might be possible to place the Ca^{2+} sparks within the framework of the sarcomeres. This was attempted in the experiment illustrated in Fig. 6. The three consecutive frames in the upper row (#10, #11, and #12) show two sparks. The one in the lower left corner is unusual in being, quite strong, yet visible in all three frames thus spanning an interval of 16 ms. The other near the right side of frames #11 and #12, is very strong, and was found in a location which produced numerous spikes during repeated recording of 50-frame segments. Each of the next two rows shows a single spark appearing in one frame (#19 and #23) disintegrating or fading in the next (#20 and #24). The image in the lower right corner shows the most intense central regions of these four sparks (white) superimposed on the enhanced sarcomere pattern. The release sites appear to coincide fairly accurately with the ridges which again are thought to coincide with the z-lines. It may be noticed that the hyperactive site with repeating releases was found near the edge of the cell (dashed line).

7. ADVANTAGES AND LIMITATIONS OF CONFOCAL MEASUREMENTS OF INTRACELLULAR CALCIUM

The spontaneous Ca^{2+} waves in rat ventricular myocytes spread slowly, and their general organization can therefore be observed without making full use of the speed granted by our rapidly scanning confocal microscope (Noran). While our observation are in general agreement with previous reports, it is of interest to notice that there may be additional information to be gained by exploring the properties of the Ca^{2+} waves in relation to the cellular architecture. Thus, the spiral wave shown in Fig. 1 is similar to those reported by Lipp and Niggli (1993) but differs in the respect that it was recorded in a very wide cell. This makes it possible to clearly see the wave fronts as curved and it yields the observation that lateral spread is slower than longitudinal spread. Possibly this is due either to the filamentous nature of the contractile apparatus giving different tortuosity factors for longtudinal and transverse diffusion of Ca^{2+} or to organelles, such as mitochondria and SR, forming diffusion barriers between myofibrils. A more detailed investigation might reveal the detailed structure of the epicenter of the spiral waves, and the detailed spread of Ca^{2+} waves around the nuclei and past faults in the sarcomere pattern. This might reveal how rapidly released Ca^{2+} reaches troponin on the thick filament and to what extent individual sarcomeres along a myofibril act as independent units.

The loading of the cell via the pipette showed that the longitudinal diffusion constant for Fluo-3 is about 2.5×10^{-7} cm^2/sec (Fig. 2). Similar values for longitudinal diffusion were found by measuring the redistribution of Indo-1 and Fura-2 after bleaching one end of a cell (Blatter & Wier, 1990). A measurement of lateral diffusion might show if it were slowed to an extend which explains the slower spread of wave fronts in the lateral direction.

The confocal images presented here were recorded at a rate of one pixel every 100 ns. This is about one order of magnitude faster than the rate used in most previous studies and is utilized to its best advantage by having an acousto-optical deflector sweep the beam rapidly along one of two coordinate axis of the image plane. Presently this technique (Noran) uses a slit instead of a pinhole on the detection side and might therefore expected to be associated with some loss of confocality in the rapidly scanned direction (left-right in the present illustrations). It is important therefore to notice, that the Ca-sparks, as they first appear are small and well defined and have nearly spherical symmetry (Fig. 5).

8. TIME COURSE OF LOCAL Ca^{2+} RELEASES

The present measurements of Ca^{2+} sparks supports the notion (Adachi-Akahane *et al.*, 1996) that the release process is not altered significantly by large concentrations of Ca^{2+} buffer (1 mM Fluo-3 plus 1 mM EGTA in Fig 4; 1 mM Fluo-3 plus 5 mM EGTA in Figs. 5 and 6). The decay of the sparks, on the other hand, appears to be accelerated by the two-buffer system. Thus we find only small remnants of sparks after 8 ms (Figs 5. and 6.) while Cheng *et al* (1993) measured a half time of 25 ms in the line scan mode using Fluo-3-AM. The idea of using a two-buffer system was to measure rapid changes in $[Ca^{2+}]_i$ with Fluo-3 while slower changes were quenched by the non-fluorescent, slower acting EGTA. The brief duration of the measured sparks suggest that the approach is successful, but higher frame rates and model calculation should be used to validate and optimize the technique.

The recordings at 120 frames per second indicate that the release typically lasts less than 8 ms. Releases of similar duration may be implied from the Ca_i-transients which continue to rise for 10 ms when the Ca^{2+} current is terminated a few ms after its activation (Cleemann and Morad, 1991; Cannel *et al.*, 1995).

The decay of the Ca^{2+} sparks appear to be governed, not simply by the diffusion limited spread of a Gaussian distribution, but by an unexpected rapid return to resting levels near the center of the release (Figs 5 and 6). A possible explanation is that the release from a specific site may be followed rapidly by the reabsorption by the SR in the same region. This may imply that the resequestration by Ca-ATPase is stimulated, not so much by elevation of intracellular Ca^{2+}, as by the emptying of the SR. In this context it may be noted that the SR continues to load effectively in cells where the global intracellular Ca^{2+} activity is reduced to 20 to 30 nM (Adachi-Akahane *et al.*, 1996).

9. AMPLITUDE OF THE Ca^{2+} SPARKS

The size of the Ca^{2+} sparks may be roughly estimated from the observation that the fluorescence intensity in a 3 μm^3 region (3.8 μm^2 area, 0.8 μm depth of focal plane) transiently increases two-fold in cells equilibrated about 75% with 1 mM Fluo-3 (K$_d$ = 300 nM) and with resting Ca^{2+} activity is thought to be about 40 nM. The resulting release is 3 μ^3 × 0.75 × 1 mM × (2–1) × (40 nM/300 nM) = 10^{-15} liter × 10^{-3} M × 0.3 ≅ 3 × 10^{-19} moles. A more accurate value might be obtained by directly measuring the resting Ca^{2+} concentration and the point-spread function of the confocal instrument. Nevertheless, the value is close to 2 × 10^{-19} moles found by Cheng *et al.* (1993) who considered the capacity of native buffers in cells with less Ca^{2+} indicator dye. These authors pointed out that a release of this magnitude might result from the opening of a single ryanodine receptor for 10 ms. Yet the release may also be compared to the amount of Ca^{2+} required to activate a "sarcomeric unit". During activation the amount of Ca^{2+} in the cytosol increase by 141±18 μM in voltage clamped cardiomyocyte (Adachi-Akahane *et al.* 1996). In each "sarcomeric unit" of 1 μ × 1μ × 2μ this corresponds to a release of 2 × 10^{-15} liter × 141 × 10^{-6} M = 2.8 × 10^{-19} moles. Therefore it seems equally valid to relate each Ca^{2+} sparks to the total release from one or more t-SR junctions. On the other hand, the size of the Ca^{2+} sparks (Figs. 5 and 6) and their relationship to the sarcomere pattern (Fig. 6) indicate that they do not represent aborted waves. In fact, there was no tendency for sparks to spread from one z-line to the next, and lateral spread over submicron distances is also unlikely when considering that waves in this direction spread slower than in the longitudinal direction.

These results show Ca^{2+} sparks occur in Ca^{2+} buffered cells and can be visualized as developing and decaying in frames recorded at 120 Hz. The measurement in plane, as opposed to a line-scan, helps to determine the exact location of sparks and to distinguish between attenuated releases in one spot and releases which are slightly removed. The ability to record at even higher frame rates (240 Hz, 480 Hz) may prove useful in measuring the distribution governing the magnitude and duration of individual sparks, including the rate of occurrence of releases which are but a fraction of 10^{-19} moles (60,000 ions) of Ca^{2+}.

10. SUMMARY

Rapid confocal fluorescence microscopy was used to measure "Ca^{2+} sparks" in rat ventricular cells dialyzed with millimolar concentrations of Fluo-3 and EGTA and held at potentials from −60 to −90 mV. At a frame frequency of 120 Hz "sparks" occurred randomly near the z-line as local, roughly circular (1 to 2 μm in diameter) rises in [Ca^{2+}]$_i$ and faded away gradually in the following one or two frames. The amount of Ca^{2+} in a spark (about 3 × 10^{-19} moles) is sufficient to activate one sarcomere (2 μm) of a myofibril (1 μm in diameter) and is therefore unlikely to represent the release from a unit smaller than an entire dyadic t-SR junction. The accuracy of the measurements (< 10^{-19} moles, <0.3 μm) suggest that it may be possible to identify individual sites where Ca^{2+} release occurs repeatedly and to distinguish releases of different magnitudes and duration.

11. ACKNOWLEDGMENT

This work was supported by NIH grant ROL16152.

12. REFERENCES

Adachi-Akahane, S., Cleemann, L., & Morad, M. (1996). Cross-signaling between L-type Ca^{2+} channel and ryanodine receptors in rat ventricular myocytes. *Journal of General Physiology* (in press).

Blatter, L. A., & Wier,W.G. (1990). Intracellular diffusion, binding, and compartmentalization of fluorescent calcium indicators indo-1 and fura-2. *Biophysical Journal*, **58**:1491–1499.

Cannell, M.B., Cheng, H., & Lederer, W.L. (1995). The control of calcium release in heart muscle. *Science*, **268**:1045–1049.

Carl, S.L., Felix, K., Caswell, A.H., Brandt, N.R., Ball, W.J. Jr., Vaghy, P.L., Meissner, G., & Ferguson, D.G. (1995). Immunolocalization of sarcolemmal dihydropyridine receptor and sarcoplasmic reticulum triadin and ryanodine receptor in rabbit ventricular and atrial cells. *Journal of Cell Biology*, **129**:673–682.

Cheng, H., Lederer, W.J., & Cannell, M.B. (1993). Calcium sparks: elementary events underlying excitation-contraction coupling in heart muscle. *Science*, **262**:740–744.

Cleemann, L., & Morad, M. (1991). Role of Ca^{2+} channel in cardiac excitation-contraction coupling in the rat: Evidence from Ca^{2+} transients and contraction. *Journal of Physiology*. **432**:283–312.

Fabiato, A. (1983). Calcium-induced release of calcium from the cardiac sarcoplasmic reticulum. *American Journal of Physiology*, **245**:C1–C14.

Fabiato, A. (1985). Time and calcium dependence of activation and inactivation of calcium-induced release of calcium from sarcoplasmic reticulum of a skinned cardiac Purkinje cell. *Journal of General Physiology*, **85**:247–289.

Hamill, O.P., Marty, A., Neher, E., Sakmann, B., & Sigworth, F.J. (1981). Improved patch-clamp technique for high resolution current recording from cells and cell-free membrane patches. *Pflügers Archiv*, **191**:85–100.

Jorgensen, A.O., Shen, A.C.Y., Wayne, A., McPherson, P.S., & Campbell, K.P. (1993). The Ca^{2+}-release channel/ryanodine receptor is located in junctional and corbular sarcoplasmic reticulum in cardiac muscle. *Journal of Cell Biology*, **120**:969–980.

Lipp, P., & Niggli, E. (1993). Microscopic spiral waves reveal positive feedback in subcellular calcium signaling. *Biophysical Journal*, **65**:2272–2276.

Lopez-Lopez, J.R., Shacklock, P.S., Balke, C.W., & Wier, W.G. (1994). Local, stochastic release of Ca^{2+} in voltage-clamped rat heart cells: visualization with confocal microscopy. *Journal of Physiology*, **480**:21–29.

Näbauer, M., Callewaert, G., Cleemann, L., & Morad, M. (1989). Regulation of calcium release is gated by calcium current, not gating charge in cardiac myocytes. *Science*, **244**:800–803.

Niggli, E., & Lipp, P. (1992). Spatially restricted Ca^{2+}-release in cardiac myocytes revealed by confocal microscopy. *Pflügers Archiv*, **420**:suppl. 1, R81.

O'Malley, D.M. (1994). Calcium permeability of the neuronal nuclear envelope: Evaluation using confocal volumes and intracellular perfusion. *Journal of Neuroscience*, **14**:5741–5758.

Santana, L.F., Cheng, H., Gomez, A.M., Cannell, M.B., & Lederer, W.J. (1996). Relation between the sarcolemmal Ca^{2+} current and Ca^{2+} sparks and local control theories for cardiac excitation-contraction coupling. *Circulation Research*, **78**:166–171.

Sham, J. S. K., Cleemann, L., & Morad, M. (1995). Functional coupling of Ca^{2+} channels and ryanodine receptors in cardiac muscle. *PNAS*, **92**:121–125.

Shacklock, P.S., Wier, W.G., & Balke, C.W. (1995). Local Ca^{2+} transient (Ca^{2+} sparks) originate at transverse tubules in rat heart cells. *Journal of Physiology*, **487**:601–608.

Smith, G.D., Wagner, J., & Keizer, J. (1996). Validity of rapid buffering approximation near a point source of calcium ions. *Biophysical Journal*, **70**:2527–2539.

Stern, M.D. (1992). Buffering of calcium in the vicinity of a channel pore. *Cell Calcium*, **13**:183–192.

Sun, X.H., Protasi, P., Takahashi, M., Takeshima, H., Ferguson, D.G., & Franzini-Armstrong, C. (1995). Molecular architecture of membranes involved in excitation-contraction coupling of cardiac muscle. *Journal of Cell Biology*, **129**:659–671.

4

CALCIUM CHANNEL DIVERSITY AT THE VERTEBRATE NEUROMUSCULAR JUNCTION

Osvaldo D. Uchitel[1] and Eleonora Katz[2]

[1]Instituto de Biología Celular y Neurociencias "Prof. Eduardo De Robertis"
Facultad de Medicina
Universidad de Buenos Aires
Paraguay 2155, Buenos Aires (1121), Argentina
[2]Departamento de Biología
Facultad de Ciencias Exactas y Naturales
Universidad de Buenos Aires
Ciudad Universitaria, Buenos Aires (1428), Argentina

1. INTRODUCTION

1.1. Calcium Channel Diversity

Calcium enters the cytoplasm mainly via voltage activated calcium channels (VACC) and represents a key step in the regulation of a variety of cellular processes such as cellular excitability, neurotransmitter release, intracellular metabolism and gene expression.

Advances in the fields of molecular biology, pharmacology and electrophysiology have led to the identification of a diverse array of VACC subtypes (see Table 1). In addition to possessing distinctive structural and functional characteristics, many of these subtypes exhibit differential sensitivities to pharmacological targets.

VACC share a common set of structural elements. They are heteromeric protein complexes composed of the pore-forming α_1 subunit and associated β and disulfide-linked $\alpha_2\delta$ subunits. An additional γ subunit was found in skeletal muscle (for extensive reviews see Hofmann et al., 1994; Caterall, W.A., 1995).

The α_1 subunit constitutes the pore through which Ca^{2+} ions flow into the cell when it opens upon depolarization. It consists of a single, long chained protein with four homologous repeating units each containing six membrane spanning regions. In analogy with Na^+ channels, gating of VACC is thought to be associated with the membrane spanning S4 segment of the α_1 subunit which bears highly conserved positively charged amino acid residues. High Ca^{2+} selectivity is provided by a spiral staircase of four negatively charged glutamate residues contributed by each domain facing the transmembrane region of the pore (Tang et al., 1993). These regions are highly conserved between the Ca^{2+} channel subtypes. The α_1 subunits also contain binding sites for the Ca^{2+} channel blockers.

Calcium and Cellular Metabolism: Transport and Regulation, edited by Sotelo and Benech.
Plenum Press, New York, 1997

Table 1. Neuronal calcium channels

Channel types	Pore-forming α-1 subunits	Pharmacology (blockers)
Low voltage activated		
T	not cloned	amiloride, Ni^{2+} (low conc.)
High voltage activated		
L	α_{1C}, α_{1D}	dihydropyridines
N	α_{1B}	ω-conotoxin GVIA
P	α_{1A} (?)	FTX
		ω-agatoxin IVA (low conc.)
		ω-conotoxin MVIIC
Q	α_{1A}	FTX (?)
		ω-agatoxin IVA (high conc.)
		ω-conotoxin MVIIC
Intermediate voltage activated		
R	α_{1E} (?)	Ni^{2+} (low conc.)

Ca^{2+} channel α_1 subunits are products of at least six different genes five of which (S,C,D,A,B) have been identified. The channel proteins encoded by different cDNAs have been sequenced and functionally expressed in heterologous cell systems (see Perez-Reyes & Schneider, 1995).

Based on the amino acid sequence similarity, the α_1 subunits of Ca^{2+} channels fall into two groups. Class C and D genes encode L -type channels $\alpha_1 C$ and $\alpha_1 D$ in which the sequence are greater than 75% identical to skeletal muscle L -type $\alpha_1 S$ encoded by Class S gene. Class C gene is expressed in the heart and in other tissues while class D is expressed in neurons and in neuroendocrine cells.

The class A, B and E genes encode non L type Ca^{2+} channels which are expressed primarily in neurons. The amino acid sequences of $\alpha_1 A$, $\alpha_1 B$ and $\alpha_1 E$ Ca^{2+} channel subunits expressed by class A, B and E genes are only 25 to 40 % identical to the skeletal muscle α_1 subunit. Between them, however, the identity in the amino acid sequence is over 60%.

Although the α_1 subunits comprise the Ca^{2+} conducting pore, the auxiliary β and $\alpha_2\delta$ subunits serve important regulatory functions (Takahashi et al., 1987). The β subunits are encoded by different genes and have multiple splice variants which contribute to the molecular and functional diversity of the VACC (see Perez-Reyes & Schneider, 1995).

Based on the voltage of activation, they were classified in low voltage activated (LVA) and high voltage activated (HVA) (see Bean, 1989). Taking into consideration the inactivation rates, the single channel conductance and the sensitivity to Ca^{2+} channel agonists and antagonists, different subtypes were identified.

LVA Ca^{2+} channels are also referred to as T-type by Tsien et al (1988) because they activate at very negative potentials in a "transient" nature due to rapid inactivation. Besides, they have a tiny unitary conductance. T-type channels are involved in the generation of pacemaker activity in neurons and cardiac muscle. There is no known toxin with selective and potent effect on this type of channel. However, it has been reported to be sensitive to sFTX and to octanol, amiloride and low concentration of Ni^{2+} (Dolphin et al., 1991; Scott et al 1992).

HVA Ca^{2+} channels were originally identified by electrophysiological and pharmacological criteria. In recent years the identity of several of them were corroborated with comparative studies of their molecular structure.

Two main groups can be defined by their sensitivity to the organic compound 1–4 dihydropyridines (DHP). The DHP sensitive channels were the first HVA channels described. They were called L because of their large and long lasting currents.

L type calcium channels are widely distributed in different tissues. In the CNS antibodies directed against L type channels stained neuronal cell bodies and proximal dendrites of many central neurons (Westenbroek et al., 1992). Synaptic transmission in most parts of the brain is not affected by DHP. However, L-type as well as N-type channels seem to be involved in peptide secretion from the neurohypophysis (Lemos & Nowycky, 1989; Wang et al., 1992) and catecholamine release from chromaffin cells in the adrenal medulla.

N type Ca^{2+} channels are activated by large depolarizations. They inactivate more rapidly than L but less rapidly than T type channels. They are insensitive to DHP but are potently blocked by many toxins purified from the venom of marine snails of the *Conus* species (Olivera et al., 1994). In particular ω-conotoxin GVIA (ω-CgTx GVIA), a peptide toxin purified from the venom of the snail *Conus geographus*, blocks completely and irreversibly the N type calcium currents in neurons. In mammalian systems a high sensitivity to this toxin is a pharmacological hallmark for the presence of N-type calcium channels (Olivera et al., 1985; Regan et al., 1991).

They mediate calcium influx related to transmitter release at many CNS synapses and at the classical frog neuromuscular junction (Kerr & Yoshikami, 1984; Katz et al., 1995) but not at the mammalian neuromuscular junction (Uchitel & Protti, 1994).

The P type Ca^{2+} channel was originally described in the cell bodies of cerebellar Purkinje neuron (Llinás et al., 1989; Regan et al., 1991) hence the name P. It is a HVA channel with very slow inactivation, resistant to DHP and ω-CgTx GVIA but very sensitive to a toxin of low molecular weight (FTX) purified from the venom of the funnel web spider *Agelenopsis aperta* and to a peptide purified from the same venom, ω-AgaIVA (Llinás et al.,1989; Mintz et al., 1992).

A non L and non N type Ca^{2+} channel α_{1A} subunit cloned from mammalian brain induces in oocytes a current sensitive to the ω-AgaIVA which may represent the P type channel. However the channel in vivo is 100 fold more sensitive to the toxin than the one expressed in oocytes, suggesting that the α_{1A} subunit represents a new type of channel (Sather et al., 1993). Recently, a new type of calcium current blocked by high concentrations of ω-AgaIVA was described in the cerebellar granule cells and termed Q-type (Wheeler et al., 1994).

The proposed P and Q type channels are closely related since both are blocked by ω-AgaIVA and other toxins like the ω-conotoxin-MVIIC and they are both resistant to DHP and ω-CgTx GVIA. Their difference may arise because of heterogeneity in the α_{1A} subunit (Mori et al., 1991) or by differences in other subunits making up the channel complex.

Antibodies directed against the α_{1A} subunit (P/Q type channel) stained synaptic regions of cell bodies and dendrites and at a lower level the dendritic shaft. The antibodies are also localized with high density in presynaptic areas and in the soma and dendrites of cerebellar Purkinje neurons.

P type channels are strongly involved in transmitter release. Channels sensitive to low concentrations of ωAgaIVA and FTX mediate transmitter release at the mouse and human neuromuscular junction (Uchitel et al., 1992; Protti & Uchitel 1993; Protti et al., 1996). Synaptic transmission in most parts of the brain is not affected by DHP but is greatly inhibited by ω-AgaIVA and also by ω-Cgtx GVIA supporting the notion that multiple Ca^{2+} channels (mainly N and P/Q) are present in a single synapse and evoke exocytosis synergistically (Mintz et al., 1995).

Ca^{2+} currents unresponsive to DHP, conotoxins and agatoxins although partially blocked by Ni^{2+}, Cd^{2+} and Amiloride have been described in neurons (Regan et al., 1991; Zhang et al., 1993). These resistant currents have been attributed to a new type of channel called R-type.

A more recently cloned α_{1E} subunit induces strong Ba^{2+} currents when injected in oocytes (Soong et al., 1993; Schneider et al., 1995; Ellinor et al., 1993). Similar to the R-type Ca^{2+} currents, the α_{1E} induced currents are not affected by most specific Ca^{2+} channel blockers. The current activates at intermediate voltages and it presents a relatively fast rate of activation and inactivation (Ellinor et al., 1993). Since these currents share kinetic and pharmacological characteristics it has been suggested that α_{1E} represents an R type of channel (Zhang et al., 1993).

2. CALCIUM CHANNELS INVOLVED IN SYNAPTIC TRANSMISSION AT THE VERTEBRATE NEUROMUSCULAR JUNCTION

The identification of the type/s of calcium channels involved in the process of release at the neuromuscular junction was evaluated by studying the effects of different calcium channel blockers on synaptic transmission in mice and frogs (Uchitel et al., 1992; Protti & Uchitel, 1993; Katz et al., 1995). It had been previously known that evoked transmitter release at the mammalian neuromuscular junction was resistant to the effects of dihydropyridines, L-type VACC antagonists, and to ω-conotoxinGVIA, a potent N-type channel antagonist (Olivera, 1994), whereas evoked transmitter release at the frog neuromuscular junction was strongly antagonized by ω-CgTx but also insensitive to the action of dihydropyridines (Kerr & Yoshikami, 1984).

By means of intracellular recording in the postsynaptic muscle cell at the endplate regions, the effects of different calcium channel blockers were tested on evoked transmitter release. Figure 1 shows that in mice, the quantal content (m) of the evoked endplate potential (EPP) is strongly reduced by FTX and ω-AgaIVA, and is not affected by ω-CgTx (Uchitel, et al., 1992; Protti & Uchitel, 1993). In frogs, however, both ω-CgTx and FTX strongly reduce evoked release while ω-AgaIVA results totally ineffective even at high concentrations and for prolonged incubation times (Katz et al., 1995). As can be observed, the two P-type channel blockers, FTX and ω-AgaIVA, are effective in suppressing evoked release in mice whereas only FTX is effective in frogs. These results clearly indicate that at the mammalian NMJ, transmitter release is triggered by Ca^{2+} entry through channels of the P type family. In frogs, however, transmitter release is antagonized by only one of the P-type channel blockers and by the N-type channel blocker. This mixed pharmacology suggests that the Ca^{2+} channels involved in transmitter release in this system are not of the classical N or P-type VACC described in mammals (for extensive review see Olivera et al., 1994).

The fact that FTX is able to antagonize evoked release in both systems in a dose-dependent manner and with the same potency (IC_{50} = 16 nl ml^{-1} and 20 nl ml^{-1}, in mice and frogs, respectively) and given that ω-AgaIVA is only effective in mice whereas ω-CgTx is only effective in frogs, the effects of the toxins on Ca^{2+} entry to presynaptic terminals were evaluated by a more direct approach.

At the neuromuscular junction of vertebrate systems, the biophysical characterization of the calcium channels is hampered by the small size and inaccessibility of the synaptic terminal, which so far have precluded direct measurements of the Ca^{2+} currents

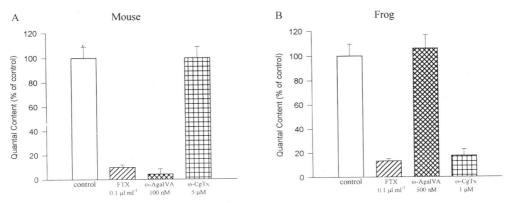

Figure 1. Effects of ω-CgTx, FTX and ω-AgaIVA on evoked transmitter release. The bar diagrams illustrate the effects of the toxins on the quantal content (*m*) of the evoked response in mice (A) and frogs (B). The preparations were incubated in either normal mouse Ringer's (phrenic-diaphragm of mice) or normal frog Ringer's (cutaneous pectoris of frogs) solution, in the presence of d-tubocurarine (d-Tc) to block muscle contraction. Quantal content was evaluated in the absence of VACC blockers (control in mice, $m = 83.3 \pm 3.5$; control in frogs: $m = 174 \pm 16$, mean ± S.E.M.). Quantal content in the presence of VACC blockers was evaluated after 1 h of incubation with either toxin and is expressed as percentage of its own control value. Each bar represents the mean ± S.E.M. of data pooled from 2–5 nerve-muscle preparations. For experimental details see Uchitel et al., 1992; Protti & Uchitel, 1993; Katz et al., 1995.

through the channels. However, by means of the perineurial technique it is possible to study the changes in conductance that occur at the synaptic terminals upon action potential arrival (Brigant & Mallart, 1982; Mallart, 1985). The extracellular recordings were obtained with a microelectrode filled with 2M NaCl placed inside the perineurial space of small nerve branches near the endplates (see Fig. 2 A). The passively propagating signals recorded at this location in the perineurium are not exactly "membrane currents" but are closely related to membrane conductance changes generated both at the terminal nodes of Ranvier and at the synaptic terminals upon stimulation of the nerve (Mallart, 1985). At this recording site, the currents generated at the terminals are picked up with reversed polarity. The first positive deflection after the stimulation artifact represents the passive discharge of the terminal membrane in response to currents associated with the propagating action potential (arrow in figure 2 D). The first negative wave corresponds to Na^+ entering the last nodes of Ranvier, the second negative wave represents the sum of the K^+ outward currents through voltage-activated K^+ channels and the inward Ca^{2+} current entering the terminals as a result of depolarization (figure 2 C and D, trace a). Under this condition the Ca^{2+} current is masked by the much greater K^+ current flowing in the opposite direction. However, in the presence of K^+ channel blockers it is possible to unmask and enhance this Ca^{2+} current (figure 2 C and D, trace b). In mouse motor terminals where there is a segregated distribution of ionic channels (Na^+ channels at the nodes and proximal regions and Ca^{2+} and K^+ channels only at the distal portions of the terminals), this combination of K^+ channel blockers disclose a Ca^{2+} signal with a fast and a long lasting component (Penner & Dryer, 1986). In frogs, there is not a clear segregation of the ionic channels along the membrane of the terminal nerve branches and there are also regional differences in the density of K^+ channels (Mallart, 1984). Therefore in frogs, by using the same combination of K^+ channel blockers used in mice, it was only possible to obtain the fast component of the owtward signal (see figure 2 D, trace b), possibly due to incomplete inactivation of the

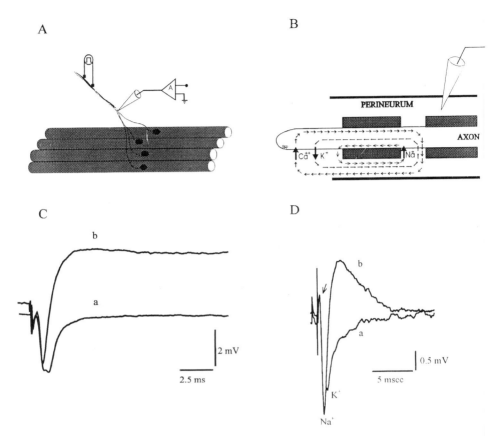

Figure 2. Perineurial currents in motor terminals. (A) schematic drawing of the levator auris nerve-muscle preparation of mice indicating the position of the intraperineurial recording electrode. (B) diagram of the terminal part
of a motor nerve fiber. Current pathways are restricted to one node for simplicity. Arrows indicate direction of ion
flow. At the recording site, the inward Na^+ currents at the last nodes of Ranvier and the outward K^+ currents generated at the terminals, flow in the same direction (Modified from Mallart, 1985). (C) presynaptic currents in mouse
motor terminals recorded in normal Ringer solution with d-Tc (30 μM) before (a) and after the application of 10
mM tetraethylammonium (TEA) and 250 μM 3,4-diaminopyridine (DAP). (D) presynaptic currents in frog motor
terminals recorded in Ringer's solution with d-Tc (30–50 μM) before (a) and after the application of 10 mM TEA
and 100 μM DAP (b).

whole population of K^+ channels. As this fast component cannot be totally attributed to
Ca^{2+}, another combination of K^+ channels blockers, which includes barium, was employed
in order to obtain the Ca^{2+} signal in which to test the different VACC blockers (Katz et al.,
1995).

In order to correlate the effects of FTX, ω-AgaIVA and ω-CgTx on the quantal content of the evoked endplate potentials with the calcium currents responsible for triggering
release, we studied their effects on the Ca^{2+} presynaptic currents (I_{Ca}). As illustrated in figure 3, both FTX and ω-AgaIVA are able to completely suppress the Ca^{2+} current in mice,
whereas ω-CgTx results ineffective in this preparation. This result is consistent with those
obtained in evoked release (Uchitel et al., 1992; Protti & Uchitel, 1993).

In the frog, the effects of ω-CgTx and ω-AgaIVA are in agreement with their effects
on evoked release. However, FTX which is able to block evoked release completely and

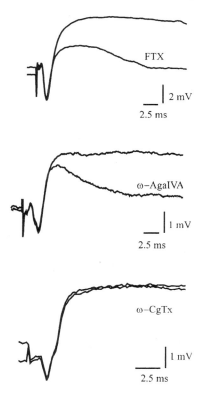

Figure 3. Effects of ω-CgTx, FTX and ω-AgaIVA on the I_{Ca} in mouse motor terminals. Control I_{Ca} were obtained in normal Ringer's solution in the presence of 30 μM d-Tc, 100 μM procaine, 10 mM TEA and 250 μM DAP. Both FTX (0.1 μl ml^{-1}) and ω-AgaIVA (100 nM) are able to abolish the Ca^{2+} component of the signal whereas ω-CgTx (5 μM) results totally ineffective. Stimulation frequency = 0.033 Hz. For experimental details see Uchitel et al., 1992; Protti & Uchitel, 1993.

with an IC_{50} similar to that found in mice, is only capable of partially blocking the I_{Ca} (Figure 4). From the incomplete blockade of a macroscopic current like the I_{Ca}, it is not possible to discriminate whether FTX is partially blocking an homogeneous population of ω-CgTx sensitive channels or totally suppressing the current through one subtype of ω-CgTx sensitive channels. The fact that FTX is able to completely suppress transmitter release at a much lower concentration than that used to block the I_{Ca} renders the possibility of partial blockade less likely (Katz et al., 1995).

Another approach to study the effects of Ca^{2+} channel blockers on presynaptic Ca^{2+} entry is to study their ability to block the calcium-activated potassium channels, gK(Ca). The activation of these channels is triggered by the increase in intracellular free Ca^{2+} subsequent to depolarization. This type of channels have been found to be present in both mammalian and amphibian synaptic terminals and can be evidenced by blocking only the "delayed rectifier" K^+ channels with DAP (Mallart, 1985b; Robitaille et al., 1993). At the frog neuromuscular junction, this gK(Ca) channels have been shown to be closely associated with the Ca^{2+} channels involved in the release process (Robitaille & Charlton, 1992; Robitaille et al., 1993), therefore we assumed that both ω-CgTx and FTX would be able to indirectly suppress this current. As shown in Fig. 5, the $I_{K(Ca)}$ is sensitive to the action of ω-CgTx but insensitive to FTX. The fact that FTX is not able to block this current gives support to the hypothesis that FTX is acting on a subpopulation of ω-CgTx sensitive channels. In the presence of a high concentration of DAP, as the one necessary to obtain the $I_{K(Ca)}$, it is possible that Ca^{2+} entering through ω-CgTx sensitive but FTX insensitive channels be enough to activate the $I_{K(Ca)}$ even in the presence of FTX. In mice,

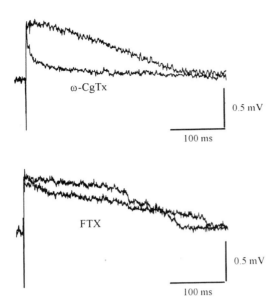

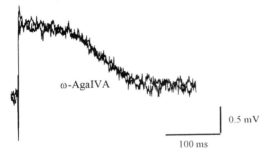

Figure 4. Effects of ω-CgTx, FTX and ω-AgaIVA on the I_{Ca} in frog motor terminals. Control I_{Ca} were obtained in normal Ringer's solution in the presence of 50 μM d-Tc, 200 μM procaine, 10 mM TEA and 200 μM $BaCl_2$. ω-CgTx (5 μM) is able to suppress most of the I_{Ca} whereas FTX (1 μl ml^{-1}) only partially suppresses this current. The recordings obtained after 40 min of incubation with 0.5 μM ω-AgaIVA show that this toxin lacks any effect on the I_{Ca}. Stimulation frequency = 0.0055 Hz. For experimental details see Katz et al., 1995.

both FTX and ω-AgaIVA are able to block the $I_{K(Ca)}$ (Protti & Uchitel, unpublished observations).

From these studies, it is clear that the population of VACC present in mouse motor terminals are mainly of the P-type family. At the frog neuromuscular junction, however, there is a major population of classical N-type VACC and a subpopulation N-type channels, the one probably involved in the release process, which is sensitive to both ω-CgTx and FTX. VACC sensitive to these two toxins have also been reported in avian preparations (Lundy et al., 1994; Alvarez-Maubecin et al., 1995). The fact that transmitter release in mice and frogs as well as in the squid giant synapse (Llinás et al., 1989), can be strongly antagonized by FTX indicates the presence of a common site in different channel subtypes sharing the same function.

From the comparative studies on the pharmacological profile of synaptic transmission in frogs and mice it becomes clear that the release of Ach at the neuromuscular junction of vertebrates can be mediated by Ca^{2+} entry through different types of VACC. Although VACC heterogeneity may possibly imply a functional diversity, the implications of the presence of one or other type of channel in relation to the process of transmitter release is still an open question.

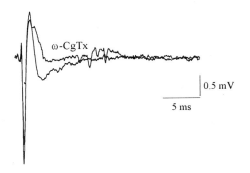

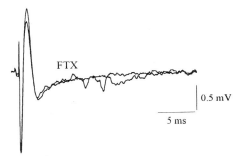

Figure 5. Effects of ω-CgTx and FTX on the $I_{K(Ca)}$ in frog motor terminals. Control $I_{K(Ca)}$ were obtained in normal Ringer's solution in the presence of 50 µM d-TC, 200 µM procaine and 500 µM DAP. ω-CgTx (5 µM) strongly depresses the $I_{K(Ca)}$ whereas FTX (1 µl ml^{-1}) results totally ineffective on this current. Stimulation frequency = 0.5 Hz. For experimental details see Katz et al., 1995.

3. REFERENCES

Alvarez, V.A., Sanchez, V.N., Rosato Siri, M.D., Cherksey, B.D, Sugimori, M., Llinás, R., & Uchitel, O.D. (1995). Pharmacological characterization of the voltage-dependent Ca2+ channels present in synaptosomes from rat and chicken Central Nervous System. J. of Neurochemistry 64: 2544–2551.

Bean, B.P. (1989). Classes of calcium channels in vertebrate cells. Annu. Rev. Physiol. 51: 367–384.

Brigant, J.L., & Mallart, A. (1982). Presynaptic currents in mouse motor endings. Journal of Physiology 333: 619–636

Catteral, W.A. (1995). Structure and function of voltage-gated ion channels. Annu. Rev. Biochem. 64: 493–531.

Dolphin, A.C., Huston, E., & Pearson, H. (1991). G protein modulation of calcium entry and transmitter release. Ann NY Acad. Sci. 635: 139–152.

Ellinor, P.T., Zhang, J.-F., & Randall (1993). Functional expression of a rapidly inactivating neuronal calcium channel. Nature. 363: 455–458.

Hofmann, F., Biel, M., Flockerzi, V. (1994). Molecular basis for Ca^{2+} channel diversity. Annu. Rev. Neurosci. 17: 399–418.

Katz, E., Ferro, P.A., & Cherksey, B.D. (1995). Effects of Ca^{2+} channel blockers on transmitter release and presynaptic currents at the frog neuromuscular junction. Journal of Physiology 486: 695–706.

Kerr, L.M., & Yoshikami, D. (1984). A venom peptide with a novel presynaptic blocking action. Nature 308:282–4.

Lemos, J.R., & Nowycky, M.C. (1989). Two types of calcium channels coexist in peptide-releasing vertebrate nerve terminals. Neuron 2:1419–26.

Llinás, R., Sugimori, M., & Lin, J.W. (1989). Blocking and isolation of a calcium channel from neurons in mammalian and cephalopods utilizing a toxin fraction (FTX) from funnel web spider poison. Proc. Natl. Acad. Sci. USA 86:1689–93.

Lundy, P.M., Hamilton, M.G. & Frew, R. (1994). Pharmacological identification of a novel Ca^{2+} channel in chick brain synaptosomes. Brain Research 643 (1–2): 204–210

Mallart, A. (1984). Presynaptic currents in frog motor endings. Pflugers Archives 400: 8–13.

Mallart, A. (1985a). Electric current flow inside perineurial sheaths of mouse motor nerves. Journal of Physiology 368: 565–575.

Mallart, A. (1985b). A calcium-activated potassium current in motor nerve terminals of the mouse. Journal of Physiology 368: 577–591

Mintz, I., Venema, V., & Swiderek, K. (1992). P-type calcium channels blocked by the spider toxin ω-Aga IVA. Nature 355:827–829.

Mintz, I.M., Sabatini, B.L., & Regehr, W.G. (1995). Calcium control of transmitter release at a cerebellar synapse. Neuron 15:675–88.

Mori, Y., Friedrich, T. & Kim, M. (1991). Primary structure and fuctional expression from complementary DNA of a brain calcium channel. Nature 350:398–402.

Olivera, B.M., Gray, W.R., & Zeikus, R. (1985). Peptide toxins from fish-hunting cone snails. Science 230:1338–43.

Olivera, B.M., Miljanich, G.P., & Ramachandran., J. (1994). Calcium channel diversity and neurotransmitter release: the omega-conotoxins and omega-agatoxins. Annu. Rev. Biochem. 63: 823–67.

Perez-Reyes, E., & Schneider, T. (1995). Molecular biology of calcium channels. Kidney International 48:1111–24.

Protti, D.A ., & Uchitel, O.D. (1993). Transmitter release and presynaptic Ca^{2+} currents blocked by the spider toxin ω-AGA-IVA. NeuroReport 5:333–6.

Protti, D.A., Reisin, R., Angelillo Mackinley, T., & Uchitel, O.D. (1996). Calcium channel blockers and transmitter release at the normal human neuromuscular junction. Neurology 46:1391–6.

Regan, L.J., Sah, D.W.Y., & Bean, B.P. (1991). Ca^{2+} channels in rat central and peripheral neurons: high threshold current resistant to dihydropiridine blockers and omega-conotoxin. Neuron 6:268–80.

Robitaille, R., & Charlton, M.P. (1992). Presynaptic calcium signals and transmitter release are modulated by calcium-activated potassium channels. The Journal of Neuroscience 12 (21): 297–305

Sather, W., Tanabe, T., & Zhang, J.F. (1993). Distinctive biophysical and pharmacological properties of class A (BI) calcium channel α_1 subunits. Neuron. 11:291–303.

Schneider, T., Wei, X., & Olcese, R. (1995). Molecular analysis and functional expression of the human type E α_1 subunits. Receptors and Channels. 2:255–70.

Scott, R.H., Sweeney, M.I., Kobrinsky, E.M. (1992). Actions of arginine polyamine on voltage and ligand-activated whole cell currents recorded from cultured neurones. Br. J. Pharmacol. 106:199–207.

Soong, T.W., Stea, A., & Hodson, C.D. (1993), Structure and functional expression of a member of the low voltage-activated calcium channel family. Science 260:1133–6.

Takahashi, M., & Catteral, W.A. (1987). Identification of an α-subunit of dihydropiridine-sensitive brain calcium channels. Science 236:88–91.

Tang, S., Mikala, G., Bahinski, A., Yatani, A., Varadi, G., & Schwartz, A. (1993). Molecular localization of ion selectivity sites within the pore of a human L-type cardiac calcium channel. J. Biol. Chem. 268:13026–9.

Tsien, R.W., Lipscombe, D., Madison, D.V. (1988). Multiple types of neuronal calcium channels and their selective modulation. Trends Neurosci. 11:431–8.

Uchitel, O.D., & Protti, D.A. (1994). P-type calcium channels and transmitter release from nerve terminals. News in Physiological Sciences 9:101–5.

Uchitel, O.D., Protti, D.A., & Sanchez, V. (1992). P-type voltage dependent calcium channel mediates presynaptic calcium influx and transmitter release in mammalian synapses. Proc. Natl. Acad. Sci. USA 87:3330–3.

Wang, X., Treistman, S.N., & Lemos, J.R. (1992). Two types of high-threshold calcium currents inhibited by ω-conotoxin in nerve terminals of rat neurohypophysis. J. of Physiol. 445:181–99.

Westenbroek, R.E., Hell, J.W., & Warner, C. (1992). Biochemical properties and subcellular distribution of N-type calcium channel α_1 subunits. Neuron 9:1099–15.

Wheeler, D.B., Randall, A., & Tsien, R.W. (1994). Roles of N-type and Q-type Ca^{2+} channels in supporting hippocampal synaptic transmission. Science 264:107–11.

Zhang, J.F., Randall, A.D., & Ellinor, P.T. (1993). Distinctive pharmacology and kinetics of cloned neuronal Ca^{2+} channels and their possible counterparts in mammalian CNS neurons. Neuropharmacology 32:1075–88.

5

COMPARISON OF THE EFFECTS OF BDM ON L-TYPE Ca CHANNELS OF CARDIAC AND SKELETAL MUSCLE

Gonzalo Ferreira, Pablo Artigas, Rafael De Armas, Gonzalo Pizarro, and Gustavo Brum

Departamento de Biofísica
Facultad de Medicina
Gral. Flores 2125, 11800 Montevideo, Uruguay

1. INTRODUCTION

Effects of the compound 2,3 Butanedione monoxime (BDM) on force development have been described in skeletal muscle (Fryer et al., 1988), cardiac muscle (Bergey et al., 1981; West & Stephenson 1989) as well as in smooth muscle (Österman et al., 1993; Watanabe, 1993). It inhibits contraction acting at different levels: on the contractile mechanism as was shown by Horiuti et al. (1988) and Österman et al. (1993) and on the excitation-contraction coupling process (Hui & Maylie, 1991; De Armas et al., 1993; Li et al., 1985). In addition to these effects on contractility the drug reduces Ca^{2+} current through L-type Ca^{2+} channels in cardiac (Coulombe et al., 1990; Chapman, 1992; Ferreira et al., 1993), skeletal muscle (Fryer et al., 1988) and smooth muscle (Lang & Paul, 1991). This reduction obeys to an enhanced voltage dependent inactivation of the channel (Chapman 1992, 1993; Ferreira et al., 1993). Since BDM is a chemical phosphatase, member of a group of oximes with the ability to reactivate cholinesterase after exposure to organophosphorous compounds (Wilson & Grinsberg, 1955), it has been suggested that dephosphorylation is the mechanism of action of the drug. Several experimental evidences recently provided are in line with this hypothesis (Chapman, 1993a; Chapman, 1995).

The main goal of the present study was to compare the effects of BDM on currents through L-type Ca^{2+} channels and charge movement in cardiac and skeletal muscle. Charge movement in cardiac muscle originates in the gating of ionic channels (Hadley & Lederer, 1988, 1991; Bean & Rios, 1989; Shirokov et al., 1992) whereas in skeletal muscle it is mainly generated by the voltage sensor of the excitation-contraction coupling (ECC) mechanism (Schneider & Chandler, 1973; for a recent review see Ríos & Pizarro, 1991). There are pharmacological and structural similarities between the L-type cardiac Ca^{2+} channel and the voltage sensor of ECC (Rios & Brum, 1987; Tanabe et al., 1987). From a functional point of view the latter is related to the activation of the Ca^{2+} release channel of the sarcoplasmic reticulum (SR). We have provided evidences that suggest that BDM probably suppresses Ca^{2+} release by

Calcium and Cellular Metabolism: Transport and Regulation, edited by Sotelo and Benech.
Plenum Press, New York, 1997

a direct action on the SR channel (De Armas et al., 1993). This effect takes place without altering charge movement significantly. On the other hand, as was cited above, it was reported that BDM reduces I_{Ca} in skeletal muscle. These data raise an old question, namely, to was extent is charge movement in skeletal muscle related to the gating of L-type Ca^{2+} channels. The comparison proposed here may add new evidences in this respect. Additonally a characterization of the effects of BDM on the skeletal muscle Ca^{2+} channel is still lacking.

2. CELLULAR PREPARATIONS AND EXPERIMENTAL PROCEDURES

Two different cellular preparations were used for the experiments. Ventricular guinea pig myocytes were isolated following the method described by Mitra & Morad (1985) and Bean and Rios (1989) with some minimal modifications. In this preparation membrane currents were recorded with the patch clamp technique in the whole cell configuration (Hammill et al 1981). Conventional patch pipettes, with resistances in the range from 1–3 Mohms when filled, were prepared with borosilicate glass (Corning 7052, Garner Glass, Claremont,CA). The pipette solution contained (in mM) CsCl 125, $MgCl_2$ 3, EGTA 10, HEPES 10 and ATPMg 5 (pH 7.5). For recording Ba^{2+} currents the external solution contained (in mM) NaCl 100; CsCl 25, $MgCl_2$ 2, $BaCl_2$ 10 and HEPES 10 (pH 7.3). In order to measure intramembrane gating currents this solution was replaced by a solution containg (in mM): CsCl 125, $MgCl_2$ 2, $LaCl_3$ 0.1–0.5, $CdCl_2$ 6 and HEPES 10 (pH 7.3). This experiments were carried out at room temperature (20–22°C).

The skeletal muscle preparation consisted of single fibers dissected from the semitendinosus muscle of the frog (Rana catesbeiana) voltage clamped in a double vaseline gap chamber following the method previously described by Kovacs et al. (1983). The solutions used during the experiments were designed to block all ionic conductances except for Ca^{2+}. The extracellular solution contained (in mM) $TEACH_3SO_3$ 130, $MgSO_4$ 8, $CaSO_4$ 10, Trismaleate 10, 3,4 diaminopyridine 1, 9-antranilic acid 1 and TTX 10^{-3} (pH 7). The solution in the endpools had the following composition (in mM) Cs Glutamate 120, Cs Trismaleate 10, MgATP 5 and EGTA 0.1 or BAPTA 10. Intracellular free Ca^{2+} was set to 50 nM or 0 when BAPTA was used. Charge movement was recorded at 10–15°C whereas I_{Ca} was measured at room temperature (20–22°C).

BDM was dissolved in the corresponding external solution at the indicated concentrations (10–20 mM). No compensation was made for osmolarity.

In both preparations membrane currents and voltage were digitized at 1–5 kHz with a 12 bit, 100 kHz A/D board (Labmaster TM-100, Scientific Solutions, Cleveland, Ohio). Averaged test and control currents were stored on disk for later processing. Ca^{2+} or Ba^{2+} currents were recorded during positive test pulses applied from the corresponding holding potential. From the total current elicited by the test pulse the current recorded during a subthreshold control pulse appropriately scaled was subtracted in order to eliminate linear capacitive and leakage currents. Asymmetric intramembrane currents were obtained with a similar procedure as Ca^{2+} currents but with control pulses recorded from a very negative or positive subtracting holding potential (s.h.p.). At very positive or negative voltages, membrane currents are linear (Brum & Rios, 1987; Shirokov et al., 1992). Controls for cardiac gating currents were obtained with 50 mV pulses applied from a s.h.p. of +30 mV. Skeletal muscle controls were 20 mV positive pulses applied from a s.h.p. of −140 mV. After subtracting the control current the records were corrected for sloping baseline (Kovacs et al., 1983; Shirokov et al., 1992). The ON and OFF transients were thereafter time integrated independently to obtain the charge moved at the ON or the OFF respectively. The computed charge was finally normalized to cell capacitance. In cardiac myocytes cell capacitance ranged from 100 to 200 pF. In skeletal muscle fibers the voltage clamped membrane had capacitances from 10 to 20 nF.

3. BDM REDUCES CHARGE MOVEMENT

Fig. 1 compares the effects of BDM on intramembrane charge movement in cardiac and skeletal muscle. Fig. 1A shows gating currents recorded in a cardiac myocyte in response to a depolarizing test pulse to +20 mV. The membrane potential in cardiac cells was routinely held at −40 mV in order to inactivate the Na^+ current. BDM produced only a small reduction on the charge moved during the ON transient. To the contrary the OFF was affected to a larger extent. The record at the bottom of Fig. 1A corresponds to the difference between test current in reference minus the test current in BDM and evidentiates the differential effect of the compound on ON and OFF transients. This effect of BDM is reversible after washout of the drug (not shown). Similar results were obtained in all myocytes tested. Q_{on} reduction was in the average 5.2±2% (n=6), whereas in the same cells Q_{off} reduction reached 40±6%. Fig. 1B shows the effects of 10 mM BDM on the charge recorded in a single frog skeletal muscle fiber in response to 100 ms pulses. BDM at low concentrations inhibits Ca^{2+} release in frog skeletal muscle fibers (Maylie and Hui, 1991; De Armas et al. 1992). The fact that the voltage sensor of EEC is similar to the L-type Ca^{2+} channel led us to study the effects of BDM on charge movement in skeletal muscle at lower doses. The records obtained before drug treatment and afterwards practically superimpose. The lower record in Fig. 1B corresponds to the difference of records on top. Only a small effect and very similar in magnitude on both Q_{on} and Q_{off} is evident. In 4 fibers studied there were no significant differences between ON and OFF charge. The effects on the maximum mobile charge evaluated during a pulse to +20 mV (average of

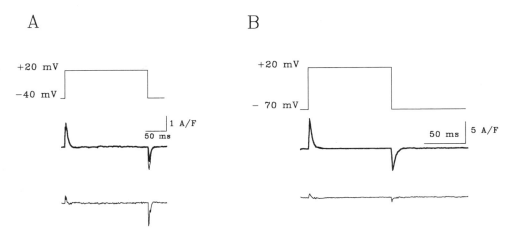

A

B

Figure 1. Comparison of the effects BDM on charge movement in cardiac and skeletal muscle. A) Superimposed are charge movement records obtained in a cardiac myocyte in reference solution (thin line) and after external application of 20 mM BDM (solid line). On top the pulse protocol used is represented. Holding potential was -40 mV. The lower trace results from the subtraction of the charge movement record in reference minus the record obtained in the presence of BDM. Cell capacitance 155 pF B) Similar experiment as in A but in a single skeletal muscle fiber. A prepulse to -70 mV was applied 50 ms before the test pulse. Charge movement records (middle traces) in reference (thin line) and 10 mM BDM (thick line) are superimposed. Bottom trace corresponds to the difference of the charge movement records. Holding potential was -80 mV.

ON and OFF) was in the average a reduction of 13.3±6% In these experiments intracellular Ca^{2+} release was virtually abolished because of the presence of 10 mM BAPTA in the internal solution. In unstreched fibers in which Ca^{2+} is being released, the charge recorded in response to small amplitude voltage pulses presents a delayed hump during the ON transient. This component is known as $Q\gamma$ (Huang, 1989) and probably depends on intracellular Ca^{2+} release (Pizarro et al., 1991). BDM drastically reduced this component in parallel with the supression of Ca^{2+} release (De Armas et al., 1993, Hui & Maylie, 1991).

4. BDM PROMOTES VOLTAGE DEPENDENT INACTIVATION OF I_{Ca} IN CARDIAC AND SKELETAL MUSCLE

As was previously shown by several authors (Coulombe et al., 1990; Chapman, 1992; Chapman, 1993; Ferreira et al., 1993) BDM enhances inactivation of I_{Ca} in cardiac cells. In order to avoid Ca^{2+} dependent inactivation of Ca^{2+} channels we replaced in our experiments external Ca^{2+} for Ba^{2+}. Fryer et al. (1988) reported that BDM also in skeletal muscle reduced the peak amplitude of the Ca^{2+} current. They used rabbit skeletal muscle fibers and perfused BDM at concentrations of 10 and 20 mM. Fig. 2 compares the effects

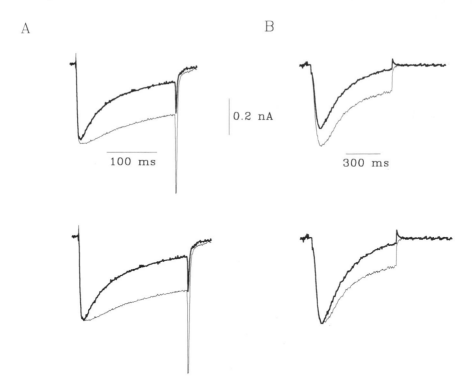

Figure 2. BDM enhances rate of inactivation of currents through L-type Ca^{2+} channel. A) Ba^{2+} current recorded in a cardiac myocyte in response to a +20 mV pulse from a holding potential of -40 mV before (thin trace) and after 20 mM BDM (thick trace). The same records scaled to match peak amplitudes are represented at the bottom. B) Ca^{2+} currents recorded from a cut skeletal muscle fiber. Pulse amplitude was +20 mV starting from a holding potential of -90 mV. Current in reference solution (thin trace) is superimposed to the current measured after perfusing 20 mM BDM (thick trace). Lower traces correspond to the same records scaled to equal peak amplitude. Note different time scale in A and B.

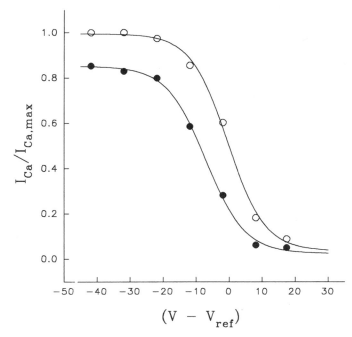

Figure 3. BDM shifts the inactivation curve of I_{Ca} in skeletal muscle. Ca^{2+} currents were recorded in response to a test pulse to +0 mV from a prepulse potential of –90 mV. The fibers were held at the different holdings for 10 minutes before test pulse application. The figure shows the average inactivation curve obtained from two fibers before (O) and after (●) 20 mM BDM perfusion. Data from individual fibers were fitted by the following Boltzmann equation:

$$I = I_{max} \cdot [\, 1 / (\, 1 + \exp(-(\,V - V^*) / k\,)\,] + C$$

Peak amplitudes were normalized to the maximum peak current in reference and voltages shifted by the value of V^* in reference for each individual fiber before averaging. Solid lines are Boltzmann fits to the averaged data. The parameters in reference are: $I_{max} = 0.96$, $V^* = -0.37$ mV, $k = 5.63$ mV and $C = 0.036$; in BDM : $I_{max} = 0.83$, $V^* = -7.04$ mV, $k = 5.93$ mV and $C = 0.024$.

of 20 mM BDM on I_{Ba} in cardiac myocytes (Fig. 2A) with the effects on frog skeletal muscle Ca^{2+} current (Fig. 2B). Peak amplitude is reduced in both cases. In the bottom part of the figure the records in BDM were scaled up to match the peaks of the currents in reference and in the presence BDM. The kinetic effects of the oxime are clear, it accelerates inactivation of the current, both in skeletal and in cardiac muscle. The steady state inactivation curve of the Ca^{2+} current in cardiac myocytes is shifted by about 5–10 mV negatively by BDM (Coulombe et al., 1990; Chapman, 1993a, Ferreira et al., 1993). We explored the steady state inactivation of the skeletal muscle Ca^{2+} current.

Fig. 3 shows the average curves obtained in reference and after application of 20 mM BDM for 2 fibers. Plotted in the figure are the peak amplitudes of the currents normalized to the maximum peak amplitude in reference. The data were fitted with a Boltzman function (see figure legend). In addition to the reduction of the peak amplitude of the current the curve is shifted approximately 7 mV to the left in the presence of BDM. There is no significative change in slope. These observations show that BDM affects, despite the differences in time scale, both cardiac and skeletal muscle Ca^{2+} currents, in a similar way.

5. LOW BDM CONCENTRATIONS HAVE PARADOXICAL EFFECTS ON I_{CA} IN SKELETAL MUSCLE

Fig. 4A shows an experiment in which the effect of 10 mM BDM was explored on I_{Ca} in a single skeletal muscle fiber. Pulses to different potentials were applied from a holding potential of −90 mV. Superimposed on the current records in reference (thin lines) are the currents recorded in the presence of the oxime (solid line). BDM reduced the peak amplitude but, surprisingly, it slows down current decay. This observation was confirmed in 3 other fibers and differs from the results obtained with higher BDM concentrations. It is well known that in skeletal muscle I_{Ca} decay depends not only on voltage dependent in-activation, but also on current amplitude. This probably reflects Ca^{2+} depletion in the T tu-bules (Almers et al., 1981).

The experimental records in Fig. 5 are presented to compare current decay in cardiac and skeletal muscle in our preparations. Peak current was reduced by holding the mem-brane at different depolarized potentials. I_{Ca} elicited by a large depolarizing pulse for dif-

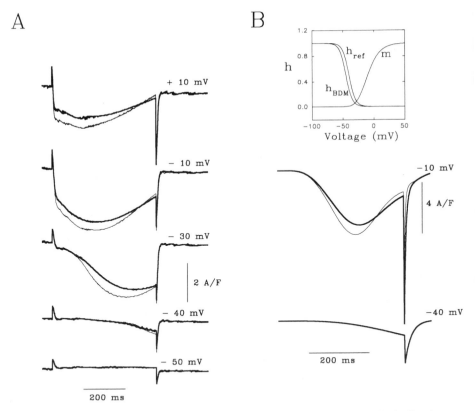

Figure 4. Effects of 10 mM BDM on I_{Ca} in skeletal muscle. A) I_{Ca} was recorded in a single fiber in response to pulses to different potentials (indicated next to each record) in reference (thin traces) and after application of 10 mM BDM (solid traces). Holding potential was -90 mV. B) Model simulations for test potentials corresponding to -10 and -40 mV. Thin traces correspond to reference, solid traces to BDM. Inset: voltage dependence of the vari-ables h and m. The effect of BDM was simulated as a left shift of the inactivation curve by 5 mV (h_{BDM}), m was assumed the same for both conditions. See text for further details of the model.

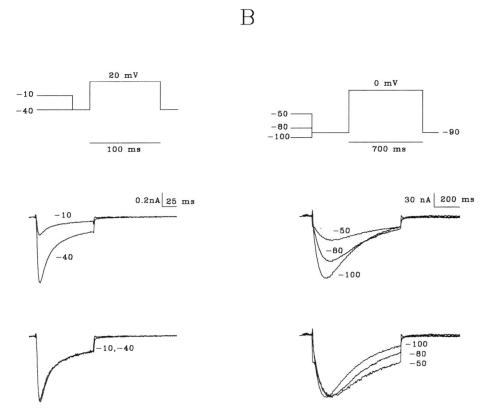

Figure 5. Dependence of the kinetics of I_{Ca} in cardiac and skeletal muscle on peak amplitude. The figure shows records of I_{Ca} obtained in response to the same test pulse but for different holding potentials in cardiac (A) and skeletal (B) muscle. On top are represented the pulse protocols. In the middle the currents corresponding to each holding potential are superimposed. In order to compare kinetics, records for each preparation were normalized to the maximum peak current and plotted superimposed. The holding potential at which the current was measured is indicated next to each experimental record.

ferent holdings (see pulse protocol on top) is plotted for cardiac (A) and skeletal (B) muscle. In order to compare the kinetics of current decay the records were scaled to make equal the peak amplitudes for each preparation. In the case of cardiac I_{Ca}, records obtained at different holding potentials perfectly superimpose after scaling (Fig. 5A bottom). This observation favors the idea that Ca^{2+}-dependent inactivation of the current depends on the local Ca^{2+} concentration in the vecinity of the channel rather than on the bulk myoplasmic Ca^{2+} concentration. In skeletal muscle on the other hand, Ca^{2+} currents differ in their kinetics showing faster decays as the peak amplitude becomes larger (Fig. 5B bottom).

6. DISCUSSION AND CONCLUSIONS

The main results presented in this study are related to the effects of BDM on charge movement and on the current through the L-type Ca^{2+} channel in cardiac and skeletal muscle. We will discuss these findings comparing the effects in both muscle types.

In cardiac muscle BDM affects charge movement reducing the amount of gating charge moved. The reduction is more prominent on the OFF transient were it reached 40% of the reference value in comparison to 5% reduction on the ON transient. This effect was a consistent finding in all myocytes tested and was reversible after washout of the drug (Ferreira et al., 1987). Gating charge immobilization is related to channel inactivation. Coulombe et al. (1990) as well as Chapman (1993) reported that the oxime augments Ca^{2+} current inactivation. In this paper we show that Ba^{2+} current through L-type Ca^{2+} channels also inactivates faster in the presence of BDM (Fig 2). A similar observation was also made by Allen et al. (1995) recently. The above mentioned findings suggest that the reduction of the OFF charge reflects an effect of BDM on voltage dependent inactivation of the channel. In skeletal muscle BDM affects charge movement to a lesser extent, the large effect on the OFF is absent in pulses up to 200 ms. We have not explored charge movement using pulses long enough to produce inactivation of skeletal Ca^{2+} current (approx. 500 ms) because in our conditions the OFF transient of such long pulses was always contaminated with ionic currents. In fig. 4 where the compound has significant effects on I_{Ca}, effects on charge movement as can be appreciated for the low voltage record are lacking. In any case this pulse is too small to promote voltage dependent inactivation of the current. For the larger pulses the OFF charge is hidden by the prominent Ca^{2+} current tail. The current records in figures 2 and 4 show that at 100 ms the effect of the oxime is very small. This probably explains the reduced effect of BDM on the OFF transient at this pulse durations. It is believed that most of the charge movement in skeletal muscle is involved in ECC. Only 5% of the dihydropyridine receptors (DHP receptors) may be functional Ca^{2+} channels (Schwarz et al., 1985). If BDM affects only these functional Ca^{2+} channels it is not expected to see large effects on the charge transients. De Armas et al. (1993) reported that BDM suppresses Ca^{2+} release in single muscle fibers without a significant effect on charge 1 (the charge associated with the active voltage sensors of ECC). But instead they reported suppression of $Q\gamma$, a component of charge movement that is probably dependent on intracellular Ca^{2+} release (Pizarro et al., 1991) and that was present in their experimental conditions.

It was previously shown that BDM promotes Ca^{2+} current inactivation in cardiac muscle. In the present paper we also show that when the current carrier is Ba^{2+} (Fig. 2) the oxime still enhances inactivation suggesting that it exerts its effects on the voltage dependent inactivation process (Kass & Sanguinetti, 1984). In skeletal muscle where Ca^{2+} dependent inactivation is absent the compound also increases current inactivation (Fig. 2B). In the same way as it shifts the steady state inactivation curve of the Ca^{2+} current in cardiac muscle (Coloumbe et al., 1990; Chapman, 1993; Ferreira et al., 1993), in frog skeletal muscle the curve is shifted to the left by a few millivolts (Fig. 3). This left shift of the curve explains the reduction in peak amplitude during drug application. When lower doses are used (10 mM) the effect on peak current becomes smaller, but instead of a faster decay of the current, inactivation turns slower. This result was unexpected taken into account the well known effect in cardiac muscle and the effect observed in skeletal muscle at higher concentrations of the oxime. A plausible explanation would reside in the fact that in skeletal muscle Ca^{2+} current decay depends on peak amplitude. Larger peaks are associated with a faster decay. This was shown in figure 5, were currents recorded at the same test potential but with different amplitudes are compared in cardiac and skeletal muscle. This effect is probably caused by depletion of the current carrier from the T tubular lumen (Almers et al., 1981). We used a simple model to simulate the experimental records. For this we assumed that the Ca^{2+} current has a Hodgkin and Huxley kinetics (m^4h) (1952). Single channel current was computed assuming a one site, two barriers Eyring model. The K_D of the site for Ca^{2+} used was 20 mM and the single channel current with the site saturated was 0.14 pA (Hess & Tsien, 1984). The site was located in the center of the electric distance, while the barriers were symmetrically positioned at 0.25 and 0.75. T tubules were assumed as an homogeneous well stirred compartment in equilibrium with the extracellular solution. According to Almers et al. (1981) the time constant of equilibration was set to 300 ms. The effect of BDM was simulated as a negative shift of the $V_h{}^*$ of inactivation (central voltage of the inactivation curve). The shift is directly related to the concentration of the oxime and affects homogeneously all the channels. This model reproduced the experimental records in Fig.4. In the same figure on the right side are plotted simulations corresponding to pulses to −10 mV and −40 mV. $V_h{}^*$ was set in this case to −40 mV in reference and −45 mV in BDM. The model is able to reproduce all the details of the experimental records. For the higher test voltage there is a clear reduction in peak amplitude and concomitantly a slower decay of the current as seen experimentally. The peak amplitude reduction is not caused by an increased steady state inactivation but by the inactivation produced by the test pulse itself. At −90 mV holding potential the level of inactivation is the same with or without BDM (See inset in figure 4 were the voltage dependence of the h variable is plotted). For the −40 mV test pulse there is little effect on the current. The simulation shows that at low concentrations of the drug were there is a minimal shift of the inactivation curve, the reduction of peak current reduces tubular depletion of Ca^{2+} thereby counteracting its contribution to the decay of the current. At higher concentrations where the shift of the inactivation curve is larger the effects on true inactivation produce a faster decay of the current. In these cases it is also difficult to estimate the contribution of tubular Ca^{2+} depletion to current decay.

Given the chemical phosphatase activity of BDM, it is believed that the mechanism of action of the compound is by desphosphorylation of the L-type Ca^{2+} channel or of a protein intimately related to it (i.e. G protein) (Trautwein & Hescheler, 1990). Several experimental evidences have been reported that support this hypothesis. Chapman (1993) observed that isoproterenol, in ATPγS loaded cells, suppressed BDM effects on Ca^{2+} current. It has been reported that changes in the level of phosphorylation modifies amplitude and rate of inactivation of the Ca^{2+} current (Yakel, 1992; Hescheler et al., 1988). In skeletal muscle Arreola et al. (1987) showed that beta-adrenergic stimulation enhances Ca^{2+} current. This effect is mediated by cAMP indicating that a similar mechanism as in heart may be operating in skeletal muscle. Brum et al. (1990) on the other hand reported that adrenaline does not modify charge movement in single frog muscle fibers. This result is not contradictory with the results in cardiac muscle, since as mentioned earlier only a small fraction of charge movement may be related to Ca^{2+} channel gating in skeletal muscle.

7. SUMMARY

The effects of 2,3 Butanedione monoxime (BDM) on calcium current and charge movement were compared in guinea pig cardiac myocytes and in frog single skeletal muscle fibers. The compound affected charge movement in both preparations. 20 mM BDM in cardiac muscle produced a large reduction of the Q_{off} transient (about 40%), whereas Q_{on} was affected to a lesser extent (5% reduction). In skeletal muscle 10 mM BDM on the other hand reduced Q_{on} and Q_{off} in a similar proportion (approximately 13%). When present, the Qγ component of skeletal muscle charge movement was suppressed. In parallel with the described effects on charge movement, the drug promoted inactivation of the calcium current (I_{Ca}) in both muscles. The steady state inactivation curve was shifted negatively 5–10 mV. The same effects on I_{Ca} were observed when extracellular Ca^{2+} was replaced by Ba^{2+} in cardiac myocytes. Low concentrations of BDM (10 mM) produced paradoxical effects in single skeletal muscle fibers: inactivation of I_{Ca} became slower while the peak amplitude was reduced.

The results reported in this paper together with previous reports by other authors indicate that BDM promotes voltage dependent inactivation of I_{Ca} in cardiac muscle. In skeletal muscle the drug promotes inactivation of I_{Ca} in a similar way. The paradoxical effects observed at low concentrations are explained if T tubule Ca^{2+} depletion is taken into account. The small effects of BDM on skeletal muscle charge movement suggest that only a small fraction of the movile charge is implicated in the gating of the slow Ca^{2+} current.

8. ACKNOWLEDGMENTS

This work was supported by grants of Proyecto de Desarrollo de Ciencias Básicas (PEDECIBA) and Comisión Sectorial de Investigación Científica de la Universidad de la República (CSIC).

9. REFERENCES

Almers, W., Fink, R. & Palade, P.T. (1981) Calcium depletion in frog muscle tubules: the decline of calcium current under maintained depolarization. Journal of Physiology 312, 177–207.

Allen, T.J.A. & Chapman, R.A. (1995) . The effect of a chemical phosphatase on single calcium channels and the inactivation of whole cell calcium current from isolated guinea-pig ventricular myocytes. Pflügers Arch. 430,68–80.

Arreola, J., Calvo, J., García, M.C. & Sánchez, J.A. (1987) Modulation of calcium channels of twitch skeletal muscle fibres of the frog by adrenaline and cyclic adenosine monophosphate. Journal of Physiology 393, 307–330

Bean, B. & Rios, E. (1989). Non-linear charge movement in the membranes of mammalian cardiac ventricular cells. Components from Na and Ca channel gating. Journal of General Physiology 94, 65–93.

Bergey, J., Reiser, J., Wiggins J. & Freeman, A. (1981). Oximes: enzymatic slow channel antagonists in canine cardiac purkinje fibres? European Journal of Pharmacology 71, 307–319.

Brum,G. & Ríos, E. (1987) Intramembrane charge movement in frog skeletal muscle fibers, properties of charge 2. Journal of Physiology 387, 489–517

Brum,G., González,S., Ferreira,G., Maggi, M. & Santi, C. (1990) Effects of adrenaline on calcium release in single fibers of frog skeletal muscle. Biophys.J. 57:342a,.

Chapman, R. (1992) The action of 2,3-butanedione monoxime (BDM), pyridine-2-aldoxime (norPAM) and pyridine-2-aldoxime methochloride (PAM) on the inactivation of the L-type calcium current in isolated guinea-pig ventricular myocytes. (Abstract) Journal of Physiology 452,196P.

Chapman, R. (1993a). The effect of oximes on the dihydropyridine-sensitive Ca current of isolated guinea-pig ventricular myocytes. Pflügers Archives 422, 325–331.

Chapman, R.A. (1995). The introduction of trypsin into the sarcoplasm of isolated guinea-pig ventricular myocytes eliminates the inhibition of the L-type Ca2+ current caused by BDM. (Abstract) Journal of Physiology 483 , 19P.

Coulombe, A; Lefevre, I., Deroubaix, E., Thuringer, D. & Coraboeuf, E. (1990). Effect of 2,3-Butanedione 2-Monoxime on slow inward and transient outward currents in rat ventricular myocytes. Journal of Molecular and Cellular Cardiology 22,921–932.

De Armas, R., González, S., Pizarro, G. & Brum, G. (1993) BDM suppresses calcium release and Q_{gamma} in skeletal muscle fibers. (Abstract) Biophysical Journal 64, 240A.

Ferreira, G., Maggi, M., Pizarro, G. & Brum, G. (1993) BDM enhances voltage dependent inactivation of L-type calcium channel in heart. (Abstract) Biophysical Journal 64, A203.

Ferreira, G., Artigas, P., Pizarro, G., & Brum, G. (1997) Butaneidione monoxime promotes voltage-dependent inactivation of L-type calcium channels in heart. Effects on gating currents. Journal of Molecular and Cellular Cardiology 29, 777–787.

Fryer, M., Neering, I. & Stephenson, D. (1988). Effects of 2,3-butanedione monoxime on the contractile activation properties of fast- and slow-twitch rat muscle fibres. Journal of Physiology 407, 53–75.

Hadley, R., & Lederer, W. (1988). Intramembrane charge movement in guinea-pig and rat ventricular myocytes. Journal of Physiology 415, 601–624.

Hadley, R., & Lederer, W. (1991). Properties of L-type calcium channel gating current in isolated guinea-pig ventricular myocytes. Journal of General Physiology 98, 265–285.

Hammill, O., Marty, A., Necher, E., Sakmann, B., & Sigworth, F. (1981). Improved patch-clamp techniques for high-resolution current recording from cells and cell-free membrane patches. Pflügers Archives 391, 85–100.

Hescheler, J., Mieskes, G., Ruegg, J., Takai, A., & Trautwein, W. (1988). Effects of a protein phosphatase inhibitor, okadaic acid, on membrane currents of isolated guinea-pig cardiac myocytes. Pflügers Archives 412, 248–252.

Hess, P., & Tsien, R.W. (1984) Mechanism of permeation through calcium channels. Nature 309, 453–456

Horiuti, K., Higuchi, H., Umazume, Y., Konishi, M., Okazaki, O. & Kurihara, S. (1988). Mechanism of action of 2,3-butanedione monoxime on contraction of frog skeletal muscle fibres. Journal of Muscle Research and Cell Motility 9,156–164.

Hodgkin, A.L., & Huxley, A.F. (1952) A quantitative description of membrane current and its application to conduction and excitation in nerve. Journal of Physiology 117, 500–544

Hui, C., & Maylie, J. (1991). Multiple actions of 2,3-butanedione monoxime on contractile activation in frog twitch fibres. Journal of Physiology 442, 527–549.

Kass, R., & Sanguinetti, C. (1984). Inactivation of calcium channel current in the calf cardiac Purkinje fiber. Journal of General Physiology 84,705–726.

Kovacs, L., Ríos, E., & Schneider, M.F. (1983) Measurement and modification of free calcium transients in frog skeletal muscle fibers by a metallochromic indicator dye. Journal of Physiology 343, 161–196.

Lang, R., & Paul, R. (1991). Effects of 2,3-butanedione monoxime on whole-cell Ca^{2+} channel currents in single cells of the guinea-pig taenia cacci. Journal of Physiology 433, 1–24.

Li, T., Sperelakis, N., Teneick, R., & Solaro, J. (1985). Effect of diacetyl monoxime on cardiac excitation-contraction coupling. Journal of Pharmacology and Experimental Therapeutics. 232,688–695.

Mitra, R., & Morad, M. (1985). A uniform enzymatic method for dissociation of myocytes from hearts and stomachs of vertebrates. American Journal of Physiology 249, 1056–1060.

Österman, A., Arner, A., & Malmqvist, U. (1993). Effects of 2,3-butanedione monoxime on activation of contraction and crossbridge kinetics in intact and chemically skinned smooth muscle fibres from guinea-pig taenia coli. Journal of Muscle Research and Cell Motility 14,186–194.

Pizarro, G., Csernoch, L., Uribe, Y., Rodriguez, M., & Ríos, E. (1991) The relationship between Qγ and Ca release from the sarcoplasmic reticulum in skeletal muscle. Journal of General Physiology 97, 913–947.

Ríos, E., & Brum, G. (1987) Involvement of dihydropyridine receptors in excitation-contraction coupling in skeletal muscle. Nature 325, 717–720.

Ríos, E., & Pizarro, G. (1991). Voltage sensor of excitation-contraction coupling in skeletal muscle. Physiological Reviews 71,849–908.

Schneider, M.F., & Chandler, W.K. (1973) Voltage dependent charge movement in skeletal muscle: a possible step in excitation-contraction coupling. Nature 242, 244–246.

Schwartz, L.M., McClesky, E.W., & Almers, W. (1985) Dihydropyridine receptors in muscle are voltage dependent but most are not functional calcium channels. Nature 314, 747–751.

Shirokov, R., Levis, R., Shirokova, N. & Ríos, E. (1992). Two classes of gating current from L-type Ca channels in guinea pig ventricular myocytes. Journal of General Physiology 99, 863–895.

Tanabe, T., Takeshima, H., Mikami, A., Flockerzi, V., Takahashi, H., Kangawa, K., Kojima, M., Matsuo, H., Hirose, T., & Numa, S. (1987) Primary structure of the receptor for calcium channel blockers from skeletal muscle. Nature 328, 313–318.

Trautwein, W., & Hescheler, J. (1990). Regulation of cardiac L-type calcium current by phosphorylation and G proteins. Annual Reviews of Physiology 52, 257–274.

Watanabe, M. (1993). Effects of 2,3-butanedione monoxime on smooth-muscle contraction of guinea-pig portal vein. Pflügers Archives 425, 462–468.

West J., & Stephenson, D. (1989). Contractile activation and the effects of 2,3-butanedione monoxime (BDM) in skinned cardiac preparations from normal and dystrophic mice (129/ReJ). Pflügers Archives 413, 546–552.

Wilson, I., & Ginsberg, S. (1955). A powerful reactivator of alkyl-phosphate inhibited cholinesterase. Biochemica et Biophysica Acta 18, 168–175.

Yakel, J. (1992). Inactivation of the Ba^{2+} current in dissociated Helix neurons: voltage dependence and the role of phosphorylation. Pflügers Archives 420, 470–478.

6

BALLS, CHAINS, AND POTASSIUM CHANNELS

Ramón Latorre,[1,2] Enrico Stefani,[3] and Ligia Toro[3]

[1]Centro de Estudios Científicos de Santiago
Casilla 16443, Santiago 9, Chile
[2]Department of Biology
Faculty of Sciences
University of Chile
Santiago, Chile
[3]Department of Anesthesiology
University of California
Los Angeles, California 90095-1778

1. INTRODUCTION

When depolarized most voltage-dependent channels undergo a process known as inactivation. This molecular rearrangement can take place in a time scale ranging from a few milliseconds to several seconds and is characterized by a decrease in ionic current with time after the onset of a depolarizing voltage pulse. The simplest kinetic scheme able to account for the workings of an inactivating ion channel is:

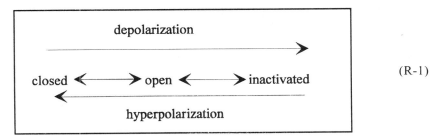

$$\text{(R-1)}$$

At present we have a detailed molecular picture of the inactivation process that undergoes a potassium channel cloned from the fruit fly (Baumann et al., 1988; Kamb et al., 1987; Tempel et al., 1987). This channel has been dubbed *Shaker* K[+] channel. *Shaker* is a *Drosophila* mutant lacking the gene that codes for the K[+] channel-forming protein. This K[+] channel belongs to a subfamily of proteins characterized by having six transmembrane domains (S1–S6). The fourth (S4) segment contains seven positively charged amino acids at every third position, with nonpolar residues in between. It has been shown unequivo-

Calcium and Cellular Metabolism: Transport and Regulation, edited by Sotelo and Benech.
Plenum Press, New York, 1997

cally that this structure is part of the voltage sensor and that it undergoes a large outwardly directed displacement during channel activation (Larsson et al., 1996; Mannuzzu et al., 1995). The S4 segment is shared by Na^+ and Ca^{2+} channels, and because of this peculiarity, the *Shaker* K^+ channel is part of a superfamily termed the S4 superfamily (Jan and Jan, 1990a). Potassium channels in this superfamily are tetramers (Liman et al., 1992; MacKinnon, 1991). Subunits belonging to the same subfamily can mix, but mixing is forbidden if the subunits originate from different genes (for reviews see Jan and Jan, 1990b; Latorre and Labarca, 1996). In this chapter we will discuss the molecular determinants of a type of inactivation process that in lies in the NH_2-terminal *Shaker* K^+ channels (N-type inactivation; Hoshi et al., 1990).

2. SEARCHING FOR BALLS AND CHAINS

Armstrong et al. (1973) found that internal perfusion of a giant squid axon with a protease (pronase) destroyed Na^+ channel inactivation. The attack by pronase is selective since the Na^+ channel activation process remains unaltered. The protease seems to destroy the channel inactivation gates in a one-by-one fashion. The conclusions of these elegant experiments were that: a) each sodium channel contains two separate gates: activation and inactivation. b) the inactivation gate is made up of protein and it is a cytoplasmic portion of the channel. Bezanilla and Armstrong (1977) proposed a specific model for the inactivation process and this is depicted in Fig. 1.

Sequence of events shown in Fig. 1 show the ball-and-chain model for Na^+ channel inactivation (Bezanilla and Armstrong, 1977). Notice that the ball and the tether (the chain) are located in the cytoplasmic portion of the channel. The ball only interacts with its receptor when the channel opens and once the ball interacts with the internal mouth (inactivated state in Fig. 1) it hinders the flow of ions and slows down the return of the activation gating charge(s). This last phenomenon is known as charge immobilization.[*] The *Shaker* K^+ channel proteins are encoded by a family of transcripts that arises by alternative splicing (Schwarz et al., 1988). The transcripts have a highly conserved central region that encompass the segments S1 to S6 and variable NH2- and COOH-terminal. The variability in the NH_2-terminal gave the first hint regarding the location of the inactivation gate. Different NH_2-terminal variants inactivate at different rates. The five known alternatively NH_2-terminal spliced variants are of different lengths and share no sequence similarity between each other (Fig. 2). Three variants (*Sh*B, *Sh*C, and *Sh*D) inactivate rapidly (Aldrich et al., 1990; Timpe et al., 1988a; Timpe et al., 1988b; Zagotta et al. 1989). On the other hand, the other two (ShD2 and ShH37) variants inactivate much more slowly (Iverson et al., 1988; Stocker et al., 1990).

Hoshi et al., (1990) found that treatment of the cytoplasmic side of the Shaker K^+ channel with trypsin destroyed inactivation. This finding demonstrates, as Armstrong et al. (1973) found in the case of the Na^+ channel, that the inactivation gate is a cytoplasmic portion of the protein. Deletions of different sizes performed in the NH_2-terminal proved that the inactivating particle is defined by the first twenty amino acids of the NH2-termi-

---

[*] Charge immobilization is measured as a decrease in the gating charge measured during repolarization (OFF gating charge). If all channels are inactivated, the rate at which the charge returns to its resting position upon repolarization is a function of the speed at which the ball leaves its receptor. Since the entire charge must be recovered in the OFF, and the recovery from inactivation is slow, the OFF charge is spread in time. Indeed, with long repolarization times the entire charge is recovered.

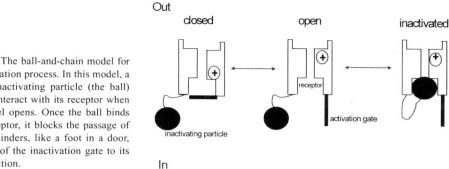

Figure 1. The ball-and-chain model for the inactivation process. In this model, a tethered inactivating particle (the ball) can only interact with its receptor when the channel opens. Once the ball binds to the receptor, it blocks the passage of ions and hinders, like a foot in a door, the return of the inactivation gate to its closed position.

nus (Fig. 2). All deletions performed inside the stretch comprised by these twenty amino acids were able to abolish completely inactivation. Even a deletion that supressed only amino acids 6 to 9 (ShBD6–9) was able to destroy the inactivation process. Deletions made outside this region *increased* the inactivation rate. On the contrary, an enlargement of the sequence lying outside the first twenty amino acids *decreased* the inactivation rate.

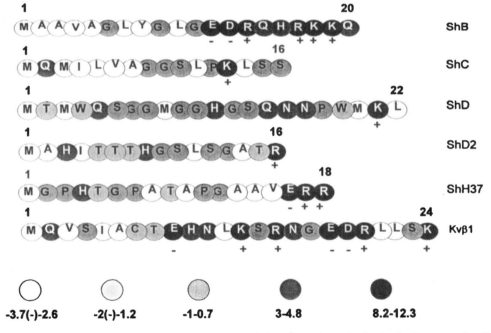

Figure 2. Primary structure of the NH2-terminal of four *Shaker* K$^+$ proteins and a β subunit of a mammalian K$^+$ channel (Kv1.1, Baumann et al., 1988; Retingg et al., 1994). The peptides corresponding to the spliced *Shaker* variants *Sh*B, *Sh*C and *Sh*O are able to induce inactivation in the K$^+$ channel *Sh*B▴6–46. Peptides *Sh*D2 and *Sh*H37 are inactive (Murrell-Lagnado and Aldrich, 1993). Different shadowing patterns and values at bottom represent the hydrophobicity of each aminoacid given as the free energy transfer for the aminoacid side chain in alpha-helical polypeptides (water oil in kcal/mol; Engelman et al., 1986). Single-letter aminoacid codes are: A, Ala; R, Arg; D, Asp; N, Asp; C, Cys; Q, Gln; H, His; I, Ileu; L, Leu; K, Lys; M, Met; F, Phe; P, Pro; S, Ser; T, Thr; Y, Tyr; W, Trp; V, Val; E, Glu; G, Gly.

Thus, it appears that the ball-and-chain model developed for the Na^+ channel is appropriate for the *Shaker* K^+ channel. The first 20 amino acids in the NH_2-terminal comprise the structure (the ball) that interacts with the internal mouth of the channel causing inactivation. Since modifications of the region comprised between the peptide that forms the ball and S1 change the rate but not the degree of inactivation. The ball is "hanging" from and connected to the rest of the protein by a chain of about 60 amino acids. This model was tested by Zagotta et al. (1990) adding a synthetic ball (ShB peptide, Fig. 2) into the internal solution bathing a membrane containing mutant *Shaker* K^+ channels that do not show N-type inactivation (ShBD6–46). Once added to the internal solution, the ball peptide is able to restore inactivation in a concentration-dependent manner. A point mutation, substitution of leucine 7 by glutamate (L7E), disrupted the peptide inactivating action. The same mutation in the NH2-terminal of the *Shaker* K^+ channel abolish inactivation (Hoshi et al., 1990). In general, disrupting the ball peptide hydrophobicity or its positive charge greatly decreased the degree of inactivation. The inactivation mechanism of a mammalian *Shaker* named Kv1.4 was also circumscribed to the NH2-terminal (Tseng-Crank et al, 1993).

3. HOW MANY BALLS ARE NEEDED TO INACTIVATE A CHANNEL?

Since K^+ channels are tetramers (Liman et al., 1992; MacKinnon, 1991), each of them must contain four NH_2-terminals and, therefore, four inactivating particles. I would like to answer the following questions: a) how many balls are needed to inactivate a K^+ channel? b) are the inactivating gates acting independently of each other? The answer to these two questions was given by MacKinnon et al. (1993) took advantage of the fact that mutated and wild type subunits can mix in the membrane. The amount of each of the species formed follows a binomial distribution. In one type of experiment, the mRNA coding for two different subunits were injected into *Xenopus* oocytes: wild type (empty circles and ball and stick in Fig. 3 top) and a subunit with a deleted inactivation gate and a mutation (D431N) that renders the channel insensitive to a scorpion toxin (filled circles in Fig. 3 top). With the proviso that one wild type subunit in the channel is enough to confer toxin sensitivity, all channels containing at least *one* ball will be toxin sensitive. Adding toxin to this mixture of channels showed that all of the channels that inactivate are wiped

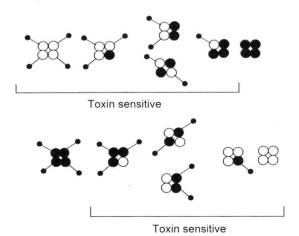

Figure 3. Mixing and matching. Top: wild type (WT) monomers (empty cycles and sticks) were mixed with subunits containing no inactivation gate and the mutation D431N. Bottom: toxin insensitive subunits containing the inactivating gate were mixed with noninactivating channels.

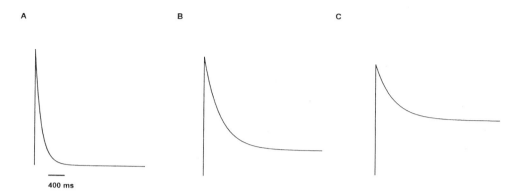

Figure 4. Simulation of current records assuming that a channel contains four inactivating particles and that the inactivation rate is proportional to the number of gates. The modelling assumes that the inactivating state is fully absorbent: closed ↔ open → inactivated. The time constant was set at 100 ms for the channel with four gates (A). In (B) channels with four gates were diluted with channels with no gates setting the non-inactivating current at 20% of the peak (Tau = 200 ms). (C) Non-inactivating current set at 50% of the peak (Tau = 300 ms).

out by the toxin suggesting that *one ball is enough to inactivate a channel.* A new twist can be introduced to this experimental approach by combining toxin insensitive-inactivating subunits with toxin sensitive noninactivating subunits. In this case, the channel distribution is like the one shown in Fig. 3 (bottom). Therefore, the toxin should block most of the current leaving a fraction of channels that inactivate fully. This prediction was corroborated experimentally. The second question was answered by diluting inactivating subunits with noninactivating subunits. When the fraction of noninactivating subunits is large, the current is mostly induced by noninactivating channels, and the small fraction that inactivates is almost entirely due to channels containing only one ball. Experimentally MacKinnon et al. (1993) found that as the wild type are diluted with noninactivating subunits, the fraction of sustained current increases and the fraction of inactivating current decreases and becomes *slower* (Fig. 4). In fact, the inactivation rate constant of channels with one ball was one-fourth that of channel with four inactivating gates. In contrast the rate of recovery from inactivation is independent of the number of gates. These results strongly suggest that *each of the four balls in a Shaker K^+ channel is independent.* Gomez-Laguna and Armstrong (1995) reached similar conclusions by destroying the *Shaker* K^+ channel inactivation with papain. They showed that as papain exerts its enzymatic action, the fraction of inactivating current inactivates at a slower rate.

4. MOLECULAR NATURE OF THE INTERACTION BETWEEN THE 'BALL' PEPTIDE AND K^+ CHANNELS

4.1. The 'Ball' Peptide Interacts with the Internal Channel Vestibule

When added to the internal side tetraethylammonium (TEA) blocks *Shaker* K^+ induced currents, and produces a substantial slowing of the N-type inactivation. As pointed out by Choi et al. (1991), the simplest interpretation of this result is that TEA and the ball cannot reside in the conduction system of the channel at the same time. In fact, if the inactivating particle and TEA compete for the same site, the slowing down of the inactivation process induced by TEA can be quantitated by means of the expression:

$$t_i([TEA]) = 1/k_iPo = 1/k_i(1 + [TEA]/K_d) \qquad (1)$$

where $t_i([TEA])$ is the [TEA]-dependent inactivation time constant, k_i is the inactivation rate constant, and K_d is the dissociation constant for the reaction between the channel and TEA. We assume here that in the steady state, inactivation is complete, and that TEA blockade is fast (is in equilibrium) compared to the Open-Inactivated transition. Equation (1) can be rewritten as:

$$t_i([TEA])/to = (1 + [TEA]/K_d) \qquad (2)$$

where to = $1/k_i$. In a strictly competitive model Equation (2) is equal to the fractional increase in current Io/I(TEA) (Miller, 1988; Choi et al., 1991; Toro et al., 1992) where I(TEA) is the peak current measured in the presence of TEA. Therefore in the scheme:

$$\text{BlockedTEA + OpenInactivated} \qquad (R-2)$$

$t_i([TEA])/to$ will increase exactly by the same factor by which the peak current is reduced. As demanded by the competitive model a 2-fold reduction in peak current was accompanied by a two-fold increase in $t_i([TEA])/to$ (Choi et al., 1991). We conclude that TEA competes with the inactivating particle because they both bind to the same channel region i.e., the internal mouth (e.g. Fig. 1).

Two lines of evidence corroborate the conclusion that, as TEA, the inactivation gate behaves like an open-channel blocker (Demo and Yellen,1991; Ruppersberg et al., 1991). First, recovery from inactivation is speeded up by increasing external [K+]. This finding can be understood if K^+ moving through the pore from the external side can sweep away the ball form its site either by a knock off mechanism and/or by electrostatic reasons. Second, upon repolarization the channel reopens. See that kinetic scheme R-1 and Fig. 1 tells us that the inactivating particle must leave the pore before the channel can close.

4.2. The 'Ball' Receptor Has Been Conserved Through Evolution

In order to answer the question of whether or not the 'ball' receptor has been conserved through the evolution of K^+ channels, the blocking ability of the inactivating peptide has been tested in other K^+ channels. These include voltage-gated (Dubinsky et al. 1992; Ruppersberg et al., 1991; Zagotta et al., 1990), Ca^{2+}-activated K^+ channels (Beirao et al., 1994; Foster et al., 1992; Solaro and Lingle, 1992; Toro et al. 1992, 1994) and an ATP-dependent K^+ channel from skeletal muscle. With the exception of the Ca^{2+}-activated K^+ channel from chromaffin cells (Solaro and Lingle, 1992), these channels normally do not display inactivation. However, in most cases the inactivating peptide was able to block from the intracellular side only. K_{ATP} channels are not blocked by the ball peptide. Collins et al. (1996) have shown a sequence for a strongly inwardly rectifying K^+ channel sensitive to ATP that belongs to a different superfamily from the S4 superfamily. The predicted amino acid sequence shows only two putative transmembrane (M1-M2) domains connected by a pore region. Therefore, the 'ball' receptor appears to be conserved among at least some K^+ channels belonging to the S4 superfamily regardless of their native inactivating rate. It is interesting to note that the non-inactivating channel Kv1.1 (mammalian

Shaker) can be transformed into a rapidly inactivating current by internal addition of the *ShB* peptide or the Kv3.4 peptide (Stephens and Robertson, 1995). The Kv3.4 (mammalian *Shaw*) channel is coded by a different gene than the one that code for Kv1.1. As pointed out by Stephens and Robertson (1995), this may suggest that these peptides may have a relatively non-specific receptor site, although they are not that promiscuous since the inactivating peptide does not bind to K_{ATP} channels. Below we show that this apparent promiscuity arises as a consequence of the nature of the molecular determinants in the interaction between the peptide and its receptor.

4.3. The S4–S5 Loop Forms Part of the Receptor for the Inactivating Gate

Isacoff et al. (1991) found that mutations of five highly conserved residues localized between segments S4 and S5 altered the stability of the inactivated state in *Shaker*B K$^+$ channels (Fig. 5). In particular, mutation L385A decreases single-channel conductance *and* stabilizes the inactivated state in *Shaker*B channels. The equivalent mutation in a delayed rectifier [drk1 (Kv2.1)] stabilized the inactivated state induced in Kv2.1 by the internal application of the ball peptide and also decreased channel conductance. The Kv2.1 channel belongs to a subfamily different from *Shaker*, but the S4–S5 loop of Kv2.1 shares a high degree of identity with that of *Shaker* (Fig. 5). Mutations in the S4–S5 loop that did not affect the inactivation process also did not affect single-channel conductance. Taken together, these results are consistent with a model in which the S4–S5 loop is or is part of the receptor for the inactivating particle and forms a portion of the K$^+$ channel ion conduction machinery.

Once the ball binds to its receptor, the voltage sensor is "immobilized". The charge is not immobilized in mutant *Shaker* K$^+$ channels that lack inactivation (Bezanilla et al., 1991). These findings show that the inactivation particle interacts in some way with the voltage sensor slowing down the return of the charge to its resting position. The close proximity of the residues apparently forming part of the ball receptor with the S4 region could provide the structural basis of the coupling between activation and inactivation. In this regard, it is important to mention here that most mutations did not modify the voltage dependence of activation with the exception of L385V and L385F. L385 is close to S4 region that in voltage-gated K$^+$ channels is or forms part of the voltage sensor (Perozo et al. 1994; Papazian et al., 1995; Mannuzzu et al., 1996; Larson et al., 1996).[†]

Figure 5. Alignment of S4–S5 loop sequences for three cloned K$^+$ channels. In *Shaker* (*ShB*), mutations in positions L385, T388, S392, E395, and L396 alter the relative amplitude of the N-type inactivation. With the exception of L385 all mutations reduce the degree of inactivation. Three of these residues are conserved in the human K_{Ca} channel hslo (Wallner et al., 1995): L227, T229, S232. Kv2.1 is a delayed rectifier that belongs to a different subfamily of K$^+$ channels. It was originally dubbed drk 1 and is the mammalian counterpart of the *Drosophila shab* K$^+$ channel.

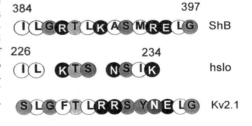

† TEA not only blocks the ionic currents but also induces charge immobilization (Bezanilla et al., 1991). This result supports the contention that the ball peptide and the hydrophobic ion interacts with similar regions of the *Shaker* K$^+$ channel (see above).

4.4. Hydrophobic and Electrostatic Interactions Are Crucial in Determining the Strength of Interaction between the Inactivating Peptide and K⁺ Channels

The structural determinants in the interaction of the *Shaker* inactivating peptide have been analyzed at the macroscopic level in Shaker K⁺ channels and at the single-channel level in Ca^{2+}-activated K⁺ (K_{Ca}) channels (Murrell-Lagnado and Aldrich, 1993a; Toro et al., 1994). In this chapter, we will discuss the structural determinants in the interaction of the peptide and a K_{Ca} channel from smooth muscle (Toro et al., 1992; Toro et al., 1994).

Internally added *Sh*B peptide (Fig. 2) blocks K_{Ca} channels in a bimolecular fashion and the inhibition can be relieved competitively by internally added TEA and by external K⁺ (Toro et al., 1992). Therefore, in non-inactivating K_{Ca} channels the *Sh*B peptide binds to the internal mouth of the channel.

Fig. 2 shows that the *Sh*B peptide can be divided into an hydrophobic region (amino acid residues 1 to 10) and a polar region containing several positive charges. Fig. 6 shows single-channel current records taken in the absence (control) and in the presence of four mutant *Sh*B peptides added to the cytoplasmic side of the membrane. The open probability for all records in the presence of the mutant peptides is about the same (0.5) implying that either changing the net charge [1+ (R17Q); 2+ (WT); 4+ (E12ND13Q)] or the hydrophobicity increases the strength of interaction of the *Sh*B peptide with the channel. Changes in the peptide charge modify only the rate constant at which the peptide enters the internal mouth (on rate) whereas the dissociation rate constant (the rate at which the peptide leaves the channel) remains unchanged. Increasing the net positive charge of the peptide or decreasing ionic strength increases the peptide association rate constant. The on rate constant vs. peptide net charge data was well fitted assuming the existence of an electrostatic potential, F, in the internal mouth of the channel (Toro et al., 1994). This potential modifies the on rate constant for the neutral peptide by a Boltzmann factor $exp(-zeF/kT)$ where z is the peptide valence. The best fit was obtained with F = -13 mV. These results suggest that the ball peptide receptor in K_{Ca} channels is adjacent to negative charges. In K_{Ca} channels the S4-S5 loop does not contain negative charges (Fig. 5). However, two glutamates are present in the S6 region of the *Drosophila* K_{Ca} channel *slowpoke* (Adelman et al. 1992) and in h*slo* (Wallner et al., 1995). Lopez et al. (1994) presented evidence that the S6 segment of *Shaker* comprises part of the internal pore vestibule. Since K_{Ca} channels are tetramers, we expect to find at least eight negative charges in this channel region.

Fig. 2 shows that there is a prevalence of hydrophobic amino acid in amino acid residues 1 to 10 in the *Shaker* B peptide. Disruption of this hydrophobic domain of the peptide promoted large changes in peptide binding. Zagotta et al. (1990) found that the L7E peptide was completely ineffective in inducing an inactivation-like process in *Shaker* K⁺ channels. The L7E peptide is also unable to block K_{Ca} channels (Foster et al., 1992;Toro et al., 1992) or a voltage-gated K⁺ channel from the basolateral membrane of enterocytes (Dubinsky et al., 1992). Fig. 6 shows that replacing alanines by valines increases the peptide blocking potency by more than two orders of magnitude. A detail kinetic analysis showed that, in contrast to the charged mutants (e.g., E12ND13Q), the A2VA3VA5V peptide does not modify the on rate, but considerably increases the mean block time (Toro et al., 1994).[‡]

[‡] There are many blocked states. It is difficult to ascribe a molecular origin to them, but based on the very high Q_{10} of the association rate constant for the reaction peptide channel, Murrell-Lagnado and Aldrich (1993b) advanced

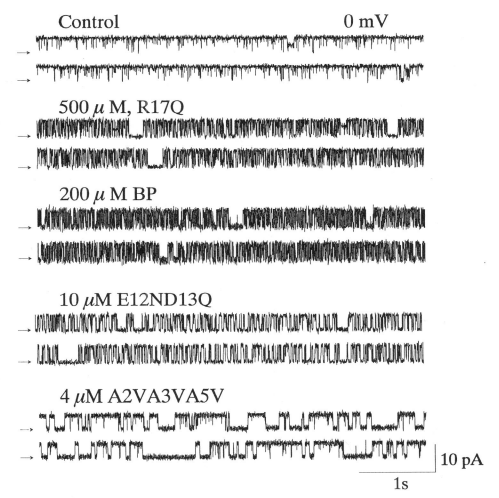

Figure 6. Increasing net charge or hydrophobicity increases peptide binding. Channel records showing channel activity before and after addition of different mutant peptides. The peptides were added to the cytoplasmic side of a K_{Ca} channel from coronary smooth muscle at concentration near peptides' Kd. The fraction of time open for each peptide: R17Q, 0.48; WT peptide, 0.49; E12QD13Q, 0.48; A2VA3VA5V, 0.5. Arrow indicates closed state. All records taken at zero mV in 250 K^+ inside and 250 Na^+ outside. (modified from Toro et al., 1994).

Thus, modifying peptide hydrophobicity primarily affects the residence time of the peptide in the internal mouth. The larger the hydrophobicity of the peptide the stronger the peptide binding to its receptor. We also tested a mutant in which two glycines were replaced by valines (G6VG9V). The overall increase in hydrophobicity of the G6VG9V peptide is the same as the A2VA3VA5V peptide with respect to the wild type peptide. In both cases the free energies differences expected from the hydrophobicity changes is

the hypothesis that they are a consequence of the presence of several different structures of the peptide in solution (See also Nobile et al., 1993). Increasing temperature increases the concentration of the peptide with the right structure for binding. A similar hypothesis was proposed by Toro et al. (1994). Fernandez-Ballester et al. (1995) showed that adoption of a β structure by the inactivating peptide is required for channel inactivation.

−3 kcal/mol. Despite this fact, the dissociation constant for the G6VG9V peptide is about 10-fold larger than that of the A2VA3VA5V peptide. This result suggests that hydrophobicity and the location in the peptide of the amino acid residues are important in determining binding strength. We have proposed the existence of an hydrophobic surface in the internal mouth of the K_{Ca} able to acommodate amino acid residues 1–8 from the N-terminus. Hence, in the case of the mutant G6VG9V, only valine 6 would contribute to the energy of transfer by interacting with the hydrophobic pocket. Residue number L7 also contributes to this energy recalling that substitution of this amino acid by glutamate dramatically decreases peptide binding (Toro et al., 1992). *We conclude that long-range electrostatic and hydrophobic interactions are crucial in determining the binding strength of a given peptide to the internal mouth of K⁺ channels.*

5. MODULATION OF INACTIVATION MEDIATED BY b SUBUNITS

Voltage-dependent K (Kv) channels are hetero-oligomeric complexes consisting of two different types of subunits: **a** subunits that are intrinsic membrane proteins with six transmembrane domains which include the molecular machinery for activation and conduction, and smaller **b** subunits. The **b** subunits are tightly associated to the cytoplasmic side of the **a** subunit (Parcej & Dolly, 1989). Rettig et al. (1994) identified cDNA clones encoding two isoforms of the β subunit of rat brain K⁺ channels. Coexpresion of one of them (Kvb1.1)° with the messenger encoding Kv1.1 leads to the expression of a rapidly inactivating current. Kv1.1 alone expresses non-inactivating currents. Similarly, Kvb1.1 speeds up by about eightfold the inactivation rate observed when Kv1.4 is expressed in isolation. Kvb1.1 subunits contains an amino terminal inactivating domain (Fig. 2) with structural similarities to the amino terminal of the **a** subunit. The Kvb1.1 has a cluster of hydrophobic amino acids and a group of charge residues. Deletion of the amino terminal of the Kvb1.1 abolishes the fast inactivation induced by this subunit when coexpressed with Kv1.1. This fast inactivation is recovered by adding the free peptide to the internal solution. The Kvb2.1 subunit that shares no homology in the amino terminal with Kvb1.1, does not alter current phenotypes in the *Xenopus* oocyte expression system (Rettig et al., 1994). Therefore, the interaction between **a** and **b** subunits may provide an inactivating "ball" to normally non-inactivating channels. Sewing et al. (1996) found that the binding site for the **b** subunit is located in the amino terminal of the **a** subunit of Kv1.5 channel. In this region of Kv1.5 there is an interaction motif (FYE/QLGE/DEAM/L) which is found only in the *Shaker*-related subfamily (Kv1). It appears that the interaction between **b** and **a** subunits does not requires of the amino terminal of the **b** subunit. The Kvb1 does not interact with other subfamilies of K⁺ channels like Kv2 and Kv3 (Nakahira et al., 1996). These studies point to a selective interaction between K⁺ channel **a**- and **b**-subunits mediated through conserved domains in the respective subunits.[+]

° This nomenclature for the K⁺ channels **b** subunits has been proposed by England et al. (1996). Cloned **b** subunits fall into three subfamilies: Kvb1.x (x = 1,2,3) have been cloned from brain and heart (England et al., 1995; Majumder et al., 1995; Morales et al., 1995; Rettig et al., 1994); Kvb2.1 was cloned from brain (Rettig et al., 1994); and Kvb3.1 was also cloned from brain (Heinenmann et al., 1995).

+ K_{Ca} channels also have a β-subunit. In this case, the β-subunit is an integral membrane protein which does not induce inactivation of the Maxi K channels, but increases its Ca^{2+}-sensitivity (Meera et al., 1996).

6. SUMMARY

The inactivation gate in *Shaker* B K⁺ channels is located in the amino terminus. Translated into the classical "ball-and-chain" model proposed by Bezanilla and Armstrong (1977) to explain Na⁺ channel inactivation, the first twenty amino acids from the amino terminus conform the ball that hangs in the following residues. One ball is enough to swing and occlude the pore once it opens. The ball hinders ion flow with the typical behavior of an open-channel blocker. The receptor for the ball appears to be located in the S4-S5 loop in K⁺ channel belonging to the S4 superfamily. This loop is highly conserved between different S4 subfamilies.

The hydrophobicity and net charge of the peptide determine the strength of interaction between the inactivating particle and the K⁺ channel. Auxiliary **b** subunits increase the rate of inactivation of specific voltage-gated K⁺ channel through a ball-and-chain mechanism. The **b** subunit binds to a site located in the amino terminal of the Kv1 **a** subunit and binding appears to be specific for this subfamily of K⁺ channels.

7. ACKNOWLEDGMENTS

This work was supported by NIH grants HL54970 (to L. Toro) and GM50550 (to E. Stefani); Chilean grants FNI 94–0227 and Catedra Presidencial and Human Frontiers in Sciences Program, European Communities Research Contract, SAREC (Sweden) and a group of Chilean private companies (CMPC, CGE, CODELCO, COPEC, MINERA ESCONDIDA, NOVAGAS, BUSINESS DESIGN ASS. and XEROX Chile) (to R. Latorre).

8. REFERENCES

Adelman, J.P., Shen, K.Z., Kavanaugh, M.P., Warren, R.A., Wu, Y.N., Lagrutta, A., Bond, C.T., & North, R.A. (1992). Calcium-activated potassium channels expressed from cloned complementary DNAs. Neuron 9:209–216.

Aldrich, R.W., Hoshi, T., & Zagotta, W.N. (1990). Differences in gating among amino-terminal variants of *Shaker* potassium channels. Cold Spring Harbor Symp.Quant.Biol. 55:19–27.

Armstrong, C.M., Bezanilla, F., & Rojas, E. (1973). Destruction of sodium conductance inactivation in squid axons perfused with pronase. J.Gen.Physiol. 62:375–391.

Baumann, A., Grupe, A., Ackermann, A., & Pongs, O. (1996). Structure of the voltage-dependent channel is highly conserved from Drosophila to vertebrate central nervous system. EMBO J. 7:2457–2463.

Beirao, P.S.L., Davies, N.W., & Stanfield, P.R. (1994). Inactivating 'ball' peptide from *Shaker* B blocks Ca²⁺-activated but not ATP-dependent K⁺ channels of rat skeletal muscle. J.Physiol.(Lond.) 474:269–274.

Bezanilla, F. & Armstrong, C.M. (1977). Inactivation of the sodium channel. I. Sodium current experiments. J.Gen.Physiol. 70:549–566.

Bezanilla, F., Perozo, E., Papazian, D.M., and Stefani, E. (1991). Molecular basis of gating charge immobilization in Shaker potassium channels. Science 254:679–683.

Choi, K.L., Aldrich, R.W., & Yellen, G. (1991). Tetraethylammonium blockade distinguishes two inactivation mechanisms in voltage-activated K+ channels. Proc.Natl.Acad.Sci.USA 88:5092–5095.

Collins, A., German, M.S., Jan, Y.-N., Jan, L.Y., & Zhao, B. (1996). A strongly inwardly rectifying K⁺ channel that is sensitive to ATP. J. Neurosci. 16:1–9.

Demo, S.D. & Yellen, G. (1991). The inactivation gate of the *Shaker* K⁺ channel behaves like an open-channel blocker. Neuron 7:743–753.

Dubinsky, W.P., Mayorga-Wark, O., & Schultz, S.G. (1992). A peptide from the *Drosophila* Shaker K⁺ channel inhibits a voltage-gated K⁺ channel in basolateral membranes of *Necturus* enterocytes. Proc.Natl.Acad.Sci.USA 89:1770–1774.

England, S.K., Uebeles, V.N., Kodali, J., Bennett, P.B., & Tamkun, M.M. (1995a). A novel K^+ channel b subunit (hKvb1.3) is produced by alternative splicing. J. Biol. Chem. 270:28531–28534.

England, S.K., Uebele, V.N., Shear, H., Kodali, J., Bennett, P.B., & Tamkun, M.M. (1995). Characterization of a voltage-gated K^+ channel b subunit expressed in human heart. Proc. Natl. Acad. Sci. USA 92:6309–6313.

Fernandez-Ballester, G., Gavillanes, F., Alvar, J.P., Criado, M., Ferragut, J.A., & Gonzales-Ros, J.M. (1995). Adoption of b structure by the inactivating "ball" peptide of the *Shaker* B potassium channel. Biophys. J. 68:858–865.

Foster, C.D., Chung, S., Zagotta, W.N., Aldrich, R.W., & Levitan, I.B. (1992). A peptide derived from the *Shaker* B K^+ channel produces short and long blocks of reconstituted Ca^{2+}-dependent K^+ channels. Neuron 9:229–236.

Gomez-Lagunas, F. & Armstrong, C.M. (1995). Inactivation in *Shaker*B K^+ channels: A test for the number of inactivating particles on each channel. Biophys. J. 68:89–95.

Heinemann, S.H., Rettig,J., & Pongs,O. (1995). Functional expression of three K channel b-subunits. Biophys. J. 68, A361 (Abstr.)

Hoshi, T., Zagotta, W.N., & Aldrich, R.W. (1990). Biophysical and molecular mechanisms of Shaker potassium channel inactivation. Science. 250:533–538.

Isacoff, E.Y., Jan, Y.N., & Jan, L.Y. (1991). Putative receptor for the cytoplasmic inactivation gate in the *Shaker* K^+ channel. Nature 353:86–90.

Iverson, L.E. & Rudy, B. (1990). The role of divergent amino and carboxyl domains on the inactivation properties of potassium channels derived from the *Shaker* gene of *Drosophila*. J. Neurosci. 10:2903–2916.

Jan, L.Y. & Jan, Y.N. (1990a). A superfamily of ion channels. Nature (London) 345:672.

Jan, L.Y. & Jan, Y.N. (1990b). How might the diversity of potassium channels be generated. Trends Neurol. Sci. 13:415–419.

Kamb, A., Tweng-Drank, J., & Tanouye, M.A. (1988). Multiple products of the *Drosophila Shaker* gene may contribute to poasssium channel diversity. Neuron 1:421–430.

Larsson, H.P., Baker, O.S., Dhillon, D.S., & Isacoff, E.Y. (1996). Transmembrane movement of the *Shaker* K^+ channel S4. Neuron 16:387–397.

Latorre, R. & Labarca, P. (1996). Potassium channels: Diversity, assembly, and differential expression. In: Potassium Channels and Their Modulators: From Synthesis to Clinical Experience, edited by Evans, J.M., Hamilton, T.C., Longman, S.D., & Stemp, G. London:Taylor & Francis, p. 123–156.

Liman, E.R., Tytgat, J., & Hess, P. (1992). Subunit stoichiometry of a mammalian K^+ channel determined by construction of multimeric cDNAs. Neuron 9:861–871.

Lopez, G.A., Jan, Y.N., & Jan, L.Y. (1994). Evidence that the S6 segment of the *Shaker* voltage-gated K^+ channel comprises part of the pore. Nature 367:179–182.

MacKinnon, R. (1991). Determination of the subunit stoichiometry of a voltage-activated potassium channel. Nature 350:232–235.

MacKinnon, R., Aldrich, R.W., & Lee, A.W. (1993). Functional stoichiometry of *Shaker* potassium channel inactivation. Science 263:757–759.

Majumder,K., DeBiasi,M., Wang,Z., & Wibble,B.A. (1995). Molecular cloning and functional expression of a novel potassium channel b-subunit from human atrium. FEBS Lett. 361:13–16

Mannuzzu, L.M., Moronne, M.M., & Isacoff, E.Y. (1996). Direct physical measure of conformational rearrangement underlying potassium channel gating. Science 271:213–216.

Meera, P., Wallner, M., Jiang, Z. & Toro, L. (1996). A calcium switch for the functional coupling between a (*hslo*) and β subunits ($K_{V,Ca}\beta$) of MaxiK channels. FEBS Lett. 382:84–88.

Miller, C. (1988). Competition for block of a Ca^{2+}-activated K^+ channel by charybdotoxin and tetraethylammonium. Neuron 1:1003–1006.

Morales, M.J., Castellino, R.C., Crews, A.L., Rasmusson, R.L., & Strauss, H.C. (1995). A novel b subunit increases rate of inactivation of specific voltage-gated potassium channel a subunit. J.Biol.Chem. 270:6272–6277.

Murrell-Lagnado, R.D. & Aldrich, R.W. (1993a). Interactions of amino terminal domains of *Shaker* K channels with a pore blocking site studied with synthetic peptides. J.Gen.Physiol. 102:949–975.

Murrell-Lagnado, R.D. & Aldrich, R.W. (1993). Energetics of *Shaker* K channels block by inactivation peptides. J.Gen.Physiol. 102:977–1003.

Nakahira, K., Shi, G., Rhodes, K.J., & Trimmer, J.S. (1996). Selective interactions of voltage-gated K^+ channels b-subunits with a-subunits. J.Biol. Chem. 271:7084–7089.

Nobile,M., Olcese,R., Chen, Y.C., Toro, L., & Stefani,E. (1993). Fast inactivation by the ball peptide in *Shaker* B channels is highly temperature dependent. Biophys. J. 64, 113a (Abstr.)

Papazian, D.M., Shao, X.M., Seoh, A., Mock, A.F., & Wainstock, D.H. (1995). Electrostatic interactions of S4 voltage sensor in Shaker K^+ channels. Neuron 14:1293–1301.

Parcej, D.N., & Dolly, J.O. (1989). Dendrotoxin receptor from bovine synaptic plasma membranes. Binding properties, purification and subunit composition of a putative constituent of certain voltage K^+ channels. Biochem. J. 257:899–903.

Perozo, E., Santacruz-Toloza, L., Stefani, E., Bezanilla, F., & Papazian, D.M. (1994). S4 mutations alter gating currents of *Shaker* K channels. Biophys. J. 66:345–354.

Rettig, J., Heinemann, S.H., Wunder, F., Lorra, C., Parcej, D.N., Dolly, J.O., & Pongs, O. (1994). Inactivation properties of voltage-gated K^+ channels altered by presence of b-subunit. Nature 369:289–294.

Roux, M. J., Toro, L., & Stefani, E. (1995). Fast inactivation of ionic currents and "charge immobilization" of *Shaker* H4 and S*h*H4 W434F K^+ channels. Biophys. J. 68:A137.

Ruppersberg, J.P., Frank, R., Pongs, O., & Stocker, M. (1991a). Cloned neuronal $I_K(A)$ channels reopen during recovery from inactivation. Nature 353:657–660.

Ruppersberg, J.P., Stocker, M., Pongs, O., Heinemann, S.H., Frank, R., & Koenen, M. (1991b). Regulation of fast inactivation of cloned mammalian $I_K(A)$ channels by cysteine oxidation. Nature 352:711–714.

Schwarz, T.L., Tempel, B.L., Papazian, D.M., Jan, Y.N., & Jan, L.Y. (1988). Multiple potassium-channel components are produced by alternative splicing at the *Shaker* locus in *Drosophila*. Nature 331:137–142.

Solaro, C.R. & Lingle, C.J. (1992). Trypsin-sensitive, rapid inactivation of a calcium-activated potassium channel. Science 257:1694–1698.

Stephens, G.J. & Robertson, B. (1995). Inactivation of the cloned potassium channel mouse Kv1.1 by the human Kv3.4 'ball' peptide and its chemical modification. J. Physiol. (London) 484:1–13.

Stocker, M., Stuhmer, W., Wittka, R., Wang, X., Muller, R., Ferrus, A., & Pongs, O. (1990). Alternative Shaker transcripts express either rapidly inactivating or noninactivating K+ channels. Proc. Natl. Acad. Sci. USA 87:8903–8907.

Tempel, B.L., Papazian, D.M., Schwarz, T.L., Jan, Y.L., & Jan, L.Y. (1987). Sequence of a probable potassium channel component encoded at *Shaker* locus of *Drosophila*. Science 237:770–775.

Timpe, L.C., Jan, Y.N., & Jan, L.Y. (1988a). Four cDNA clones from the Shaker locus of Drosophila induce kinetically distinct A-type potassium currents in Xenopus oocytes. Neuron 1:659–667.

Timpe, L.C., Schwarz, T.L., Tempel, B.L., Papazian, D.M., Jan, Y.N., & Jan, L.Y. (1988b). Expression of functional potassium channels from Shaker cDNA in Xenopus oocytes. Nature 331:143–145.

Toro, L., Ottolia, M., Stefani, E., & Latorre, R. (1994). Structural determinants in the interaction of *Shaker* Inactivating peptide and a Ca^{2+}-activated K^+ channel. Biochemistry 33:7220–7228.

Toro, L., Stefani, E., & Latorre, R. (1992). Internal blockade of a Ca^{2+}-activated K^+ channel by *Shaker* B inactivating "ball" peptide. Neuron 9:237–245.

Tseng-Crank, J., Yao, J.-A., Berman, M.F., & Tseng, G.-N. (1993). Functional role of the NH_2-terminal cytoplasmic domain of a mammalian A-type K channel. J. Gen. Physiol. 102:1057–1083.

Wallner, M., Meera, P., Ottolia, M., Kaczorowski, G., Latorre, R., Garcia, M.L., Stefani, E., & Toro, L. (1995). Cloning, expression and modulation by a b-subunit of a human maxi K_{Ca} channel cloned from human myometrium. Receptors and Channels 3:185–199.

Zagotta, W.N., Hoshi, T., & Aldrich, R.W. (1990). Restoration of inactivation in mutants of *Shaker* potassium channels by a peptide derived from ShB. Science 250:568–571.

7

THE PLASMA MEMBRANE CALCIUM PUMP

Ernesto Carafoli and Danilo Guerini

Institute of Biochemistry
Federal Institute of Technology (ETH)
8092 Zurich, Switzerland

1. INTRODUCTION

The plasma membrane Ca^{2+} ATPase has been first described in erythrocytes by Schatzmann (1966) and is now known to be present in all cells of higher eucaryotes. The pump belongs to the family of P-type ATPases (Pedersen & Carafoli, 1987a; Pedersen & Carafoli, 1987b), i.e., it forms an aspartyl-phosphate during the reaction cycle. It is a target of calmodulin (Gopinath & Vincenzi, 1977; Jarret & Penniston, 1977), which increases its affinity for Ca^{2+} by one order of magnitude, to a Kd of about 0.5 µM. The pump, however, can also be activated by a number of alternative treatments: the exposure to acidic phospholipids (Ronner et al., 1977; Niggli et al., 1981a), a controlled proteolytic treatment (Enyedi et al., 1980; Caroni & Carafoli, 1981), phosphorylations by two protein kinases, (protein kinase A (PKA) (Caroni & Carafoli, 1981) and protein kinase C (PKC) (Wright et al., 1993; Furukawa et al., 1989)), and an oligomerization process (Kosk-Kosicka & Bzdega, 1988). By general consensus, calmodulin is considered the natural modulator of the pump, but it is well to remember that the pump in the membrane is surrounded by amounts of acidic phospholipids which are, in principle, sufficient for half-maximal activation (Niggli et al., 1981b). The interaction with calmodulin has been exploited to purify the pump using calmodulin columns (Niggli et al., 1981b). The purified enzyme is active, and can be reconstituted in liposomes with optimal Ca^{2+} transport efficiency (Niggli, et al., 1981c): at variance with the Ca^{2+} pump of sarcoplasmic reticulum, whose Ca^{2+}/ATP molar transport stoichiometry is 2.0, the pump transports only one Ca^{2+} per ATP hydrolyzed.

A number of reviews on the mechanism and general properties of the pump in vivo have appeared (see for example Schatzmann, 1982; Rega & Garrahan, 1986), and the reader is referred to them for discussions of the early findings. Following the purification of the pump, numerous studies have dealt with its molecular aspects, leading to significant advances in the understanding of its structure/function relationships: recent reviews offer detailed discussions of the findings on the purified pump during the last 15 years (Carafoli, 1991; Carafoli, 1991; Carafoli, 1994; Penniston & Enyedi, 1994; Monteith & Roufogalis, 1995). This contribution will focus on the most recent developments which have emerged from the work on the purified enzyme.

Calcium and Cellular Metabolism: Transport and Regulation, edited by Sotelo and Benech.
Plenum Press, New York, 1997

2. THE TARGETING OF THE PUMP TO THE PLASMA MEMBRANE

Figure 1 shows the membrane topography of human pump isoform 4, which is the best known (the pump is the product of a multigene family). The topography is common to all other isoforms. It shows 10 transmembrane domains, and four main units protruding into the cytoplasm: the first is the N-terminal portion of the pump: its length varies between 90 and 100 residues in the various isoforms. The second (the β-strand domain) protrudes from transmembrane domain 2: its function is unknown but it contains one of the 2 binding sites for acidic phospholipids. The third unit, which is the largest, protrudes from

Figure 1. Sequence, predicted secondary structure, and membrane topology of the plasma membrane Ca²⁺ pump. CaM is the calmodulin binding domain, containing the Thr phosphorylated by PKC. PL is a basic sequence which interacts whith acidic phospholipids (the calmodulin-binding domain also interacts with them). The two black domains are the "receptors" for the autoinhibitory calmodulin binding domain. The shaded and boxed region (ATP) is the part of the pump where ATP becomes bound: it contains a conserved Lys (K). D is the site of phosphorylation. The isoform shown is PMCA4, which does not contain the C-terminal consensus site for PKA.

transmembrane domain 4: it contains the active site(s) of the pump, i.e., the site of aspartic acid phosphorylation, and the domain where ATP is bound, which includes a conserved lysine (K). The fourth unit protrudes from the last transmembrane helix and contains domains which are important for the regulation of the pump: the calmodulin binding site, consensus sites for phosphorylation by PKA and PKC (the latter site, a Thr within the cal-

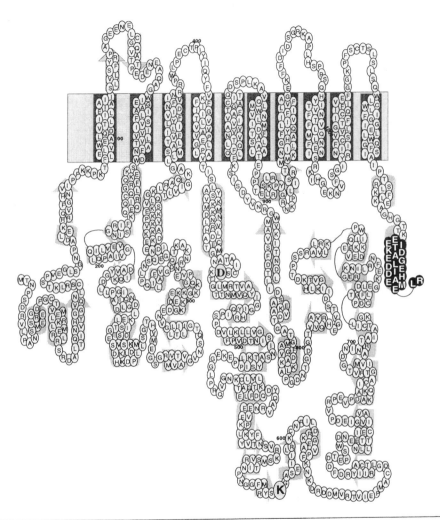

	E	E	I	T	K	D	A	E	G	L	D	E	I	D	H	A	E	M	E	L	R
PMCA△118(L,A,A,A)	-	-	-	-	-	-	-	-	-	-	-	L	-	A	-	-	A	-	A	-	-
PMCA△118(Q,Q,Q)	-	-	-	-	-	-	-	-	-	-	-	Q	-	-	-	-	Q	-	Q	-	-
PMCA△118(Q,Q)	-	-	-	-	-	-	-	-	-	-	-	Q	-	-	-	-	Q	-	-	-	-
PMCA△118(Q,N)	-	-	-	-	-	-	-	-	-	-	-	Q	-	N	-	-	-	-	-	-	-
PMCA△118(Q)	-	-	-	-	-	-	-	-	-	-	-	Q	-	-	-	-	-	-	-	-	-

Figure 2. Truncation of the pump by calpain: a C-terminal acidic sequence (black) is exposed. The inset shows the mutations that have been introduced in the C-terminal acidic sequence.

modulin binding domain, is common to all 4 gene products, the former is only present in some isoforms) and sites which bind Ca^{2+} to possibly modulate the activity of the pump (Hofmann et al., 1993). The calmodulin binding site also interacts with acidic phospholipids: as mentioned, the other phospholipid binding site is a basic stretch of about 40 amino acids located in the second cytosolic unit (Brodin et al., 1992). The second and third cytosolic units also contain sites which interact with the calmodulin binding domain. The latter interact with the main body of the pump, locks together the second and third cytosolic units, thus keeping the pump inhibited in the absence of calmodulin (Falchetto et al., 1991; Falchetto et al., 1992): calmodulin then interacts with its binding domain and swing it away from its receptor sites, relieving the inhibition.

The auto-inhibitory function of the C-terminal unit of the pump was supported by experiments on proteolysis by calpain (James, et al., 1989), which showed that the activation by the protease was linked to the truncation of the pump at the beginning of the calmodulin binding domain: attempts to mimic the effects of calpain, i.e., to produce fully active, calmodulin-independent pumps by expressing truncated versions of them led to the finding of a signal for endoplasmic reticulum (ER) retention in the C-terminal portion of the pump. A more detailed study of the membrane targeting of the PMCA pump (Zvaritch et al., 1995) showed that the cleavage by calpain exposes a very acidic sequence (Figure 2).

Truncated pumps ending with this acidic sequence were retained in the ER when expressed in Cos-7 cells (Figure 3), whereas pumps truncated about 20 residues upstream,

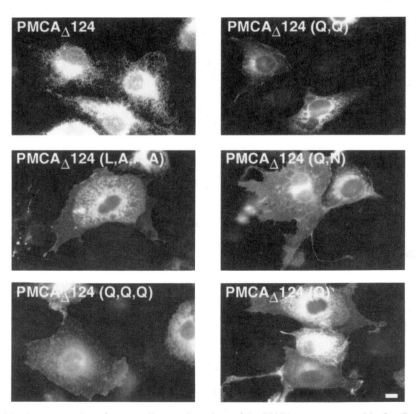

Figure 3. Membrane targeting of truncated/mutated versions of the PMCA pump expressed in Cos-7 cells. Two monoclonal antibodies, and one polyclonal antibody against the N-terminal portion of the pump have been used. (From Zvaritch et al., 1995). Bar=10µm.

and thus not containing the acidic C-terminal sequence, were correctly delivered to the plasma membrane. Point mutations to neutralise negative residues in the C-terminal sequence showed that the replacement of single acidic residues of the acidic stretch restored the correct targeting of the enzyme (Figure 3).

Since selective retention (and degradation) within the ER are common in misfolded and mutated proteins, which expose signals otherwise masked in properly folded molecules, the selective retention (and degradation) of truncated pumps in the ER could be a defense mechanism to prevent the arrival into the plasma membrane of prematurely terminated pump versions which would be permanently active.

The PMCA and the SERCA pumps are similar in functional properties and membrane topology, yet are targeted to two different membranes. To study the targeting of the two pumps, chimeras were constructed in which the N-terminal transmembrane domains of the two pumps, which are generally considered important for the targeting of membrane proteins (Machamer, 1993), were exchanged (Foletti et al., 1995). Five chimeric molecules were constructed (Figure 4): in four (chimeras A, B, C, D), different N-terminal portions of the SERCA pump were followed by the remainder of the PMCA pump molecule. In the fifth (chimera E), the N-terminal and the first two transmembrane do-

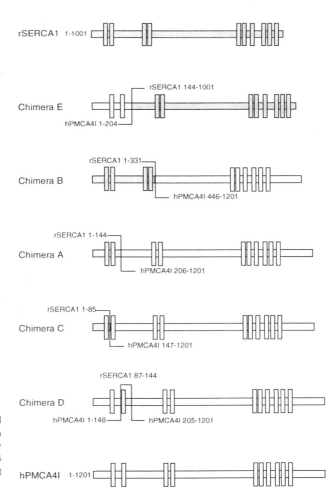

Figure 4. Chimeras of the PMCA and SERCA pumps. The isoform of each pump that has been used is shown by the abbreviations (see Carafoli, 1994 for the nomenclature). (From Foletti et al., 1995).

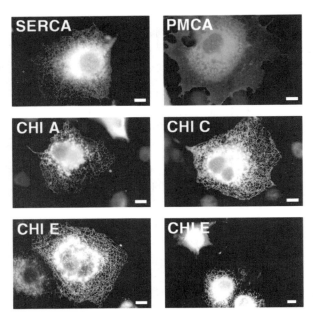

Figure 5. Membrane targeting of the PMCA/SERCA pump chimeras expressed in Cos-7 cells. Bar=30 μm. Chimera B is not shown because its targeting pattern was identical to that of chimera A. Chimera D was very rapidly proteolyzed and yielded no clear immunofluorescence pattern. Isoform-specific antibodies have been used (From Stauffer et al., 1995).

mains of the PMCA pump were followed by the remainder of the SERCA pump molecule. Surprisingly, none of the chimeras had the ability to catalyze Ca^{2+} dependent phosphoenzyme formation when expressed in Cos-7 cells: a partial exception was chimera C. Immuno-histochemistry using SERCA and PMCA pump antibodies (Figure 5) revealed that a strong ER retention signal was contained in the first 85 residues of the SERCA pump, since chimeras A-D were strictly retained in the ER (chimera D, however, was rapidly degraded). Chimera E, in which the first two transmembrane domains were PMCA type was still mostly retained in the ER, but in 5–14% of the expressing cells a plasma membrane staining pattern was visible. Thus, this chimera had lost part of the ER retention signal, i.e., that located in the first 85 amino acids. However, chimera E indicated that additional, less efficient ER retention signals exist in the SERCA pump molecule outside the N-terminus and first two transmembrane domains.

3. ISOFORMS OF THE PMCA PUMP

The four genes for the PMCA pump have been mapped to chromosomes 12, 1, 3, and X (see Carafoli & Guerini, 1993 for a recent review). Each gene produces additional isoforms by alternative splicing of primary transcripts. All four basic human gene products (Schatzmann, 1996; Pedersen & Carafoli, 1987; Gopinath & Vincenzi, 1977) have been cloned, but only 3 have so far been expressed, and studied functionally (Heim et al., 1994; Hilfiker et al. 1994; Stauffer et al. 1993).

Table 1 summarizes their most important properties, and also includes information on isoform 3, which has not been expressed as yet. Isoforms 1 and 4 are present in all tissues, and can thus be considered as products of housekeeping genes. Isoforms 2 and 3, in contrast, are only found in significant amounts in neurones: isoform 2 is typical of cerebellum, particularly of the Purkinye cells (Figure 6), and isoform 3 of the choroid

Table 1. Properties of the PMCA isoforms

	PMCA1	PMCA2	PMCA3	PMCA4
Tissue distribution	ubiquitous	restricted (brain)	restricted (brain)	ubiquitous
Level of expression in rat and human tissues	high	high	low	medium
Developmental expression/switch	isoform switch foetal/adult	isoform switch foetal/adult	down-regulated in adult tissues	isoform switch foetal/adult
KdCaM[a]	40–50 nM	8–10 nM	NA	40–50 nM
KdATP[b]	0.1 μM	0.2–0.3 μM	NA	0.7 μM
$T_{1/2}$ of the degradation by calpain	2–4 min	50–55 min	NA	50–55 min

NA = data not available.
[a]The isoform analyzed was CI, i.e., without insert at hot spot C, which encompasses the calmodulin binding domain.
[b]Determined from the phosphoenzyme formation.

plexuses. The most striking difference among isoforms is the affinity for calmodulin, which is significantly higher in isoform 2. This is interesting, considering that this isoform is typical of Purkinye cells, which have special Ca^{2+}-signalling requirements, (they also contain unusually large amounts of the inositol-tri-phosphate receptor).

A particularly interesting aspect of the isoforms is their developmental switch. Recent work on primary cultures of granular cells of rat cerebellum, which are frequently used as models for neuronal cell development, has led to some very interesting findings (Carafoli et al., 1996). The survival of these cells has a peculiar dependence on the pres-

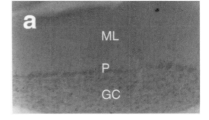

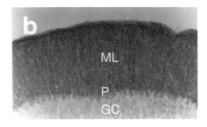

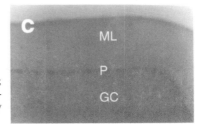

Figure 6. PMCA pump isoforms in rat cerebellum. a=PMCA1; b=PMCA2; c=PMCA3. ML=molecular layer, P= Purkinye cells (bodies), GC=granular cells. The isoform specific antibodies described by Stauffer et al. (1995) have been used.

ence of potassium or other depolarizing agents (Balas et al., 1988; Gallo et al., 1987): the treatment produces a sustained increase of cytosolic Ca^{2+}, and influences the expression of several genes for example those of plasma membrane Ca^{2+} channels, including the NMDA-glutamate receptor. Membranes from cells that had been cultured in the presence of 25 mM KCl for 1 or 9 days were probed with antibodies specific for PMCA isoforms 1, 2, 3. Isoform 4 could not be probed since the isoform 4 antibodies, raised against the human pump, do not recognise the rat enzyme (Stauffer et al., 1995). The depolarizing treatment up-regulated isoform PMCA1, most evidently in its 1CII alternatively spliced version (Fig. 5): interestingly, isoform 1CII is typical of nervous cells (Stauffer et al., 1995; Keeton et al., 1993). The most striking effects, however, were on isoforms 2 and 3 which were both dramatically upregulated. After 2 days in culture hardly any PMCA2 protein, and no PMCA3 protein at all, was detected in the cultured cells. After 5 days in the presence of 25mM potassium, the cells developed to mature neurones, and during this time, isoform 2 became very prominent, peaking at day 5–7. Isoform 3 reached its peak somewhat later, at day 7 to 9. (Fig. 7). These results suggest that the expression of the PMCA2 and PMCA3 pumps may be controlled by cytosolic Ca^{2+} whose concentration increases during the depolarization treatment. Ca^{2+} penetrates into granule cells through voltage gated, mostly L-type channels and ligand-gated, (the NMDA receptor) Ca^{2+} channels. Here, another very interesting finding was made: evidently, the path of Ca^{2+} entry into the cell matters, because inhibitors of the L-type channel blocked the up-regulation of the pump genes, whereas those of the NMDA receptor did not.

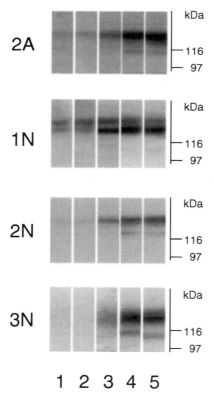

Figure 7. PMCA isoforms in granular cells of rat cerebellum. 25–30µg of crude membrane proteins from cultures of rat cerebellum granular cells incubated for 2 to 9 days in the presence of 25mM KCl were separated by SDS-PAGE electrophoresis and transferred to nitrocellulose sheets. The sheets were stained with affinity purified antibodies against the N-terminal portion of the PMCA1, PMCA2 and PMCA3 pump isoforms essentially as described by Stauffer et al, (1995). Antibody 2A recognises all 4 basic isoforms of the pump. Lane 1 = 2 days; Lane 2 = 3 days; Lane 3 = 5 days; Lane 4 = 7 days; Lane 5 = 9 days.

4. THE Ca^{2+} "CHANNEL" ACROSS THE PMCA MOLECULE

The Ca^{2+} binding sites in the C-terminal portion of the PMCA pump which were mentioned above (Hofmann et al., 1993) are likely to be only allosteric, since the removal of the C-terminal cytosolic tail of the pump, which contains them, with calpain, does not impair the ability of the enzyme to transport Ca^{2+} (James et al., 1989). The "channel" for Ca^{2+} across the molecule must by necessity require the participation of amino-acids located in the transmembrane domains. Site directed mutagenesis of selected amino acids in these domains of the SERCA pump (Clarke et al., 1989) has led to the proposal that the Ca^{2+}-"channel" is formed by six residues (four of them acidic) in transmembrane domains 4, 5, 6, and 8. These 6 residues are shown in Table 2, which aligns them with residues in equivalent positions in the PMCA pump. Some are conserved, but others are not. In particular, there is no residue equivalent to Glu 771 of transmembrane domain 5 of the SERCA pump in the corresponding domain fo the PMCA pump. Furthermore, Glu 908 in the 8th transmembrane domain of the SERCA pump is replaced by a Gln (971) in the PMCA enzyme.

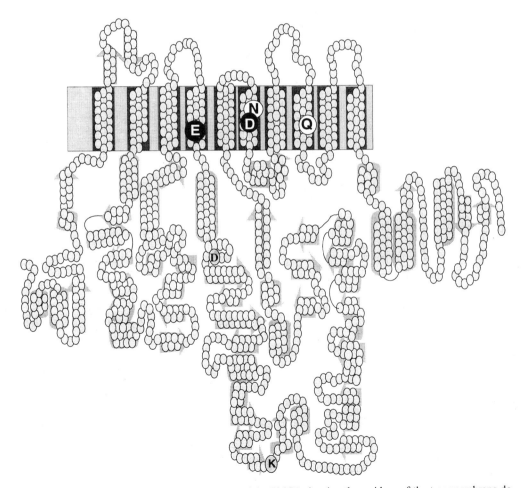

Figure 8. A scheme of the membrane topography of the PMCA showing the residues of the transmembrane domains whose mutations abolish the transport of calcium.

Table 2. Conserved residues in the transmembrane domains of the SERCA and PMCA-ATPases

	TM4	TM4	TM4	TM5	TM6	TM6	TM6	TM8
SERCA	P308	E309	P312	E771	N796	T799	D800	E908
PMCA	P422	E423	P426	A854	N879	M882	D883	Q971

Taken from Guerini et al. (1996).

The 3 conserved residues in transmembrane domains 4 and 6 and Gln 971 of domain 8 have been mutated (Guerini et al., 1996) (Figure 8). The mutations abolished the uptake of Ca^{2+} by the pump expressed in Cos-7 cells, a result which would support the conclusions reached from the mutagenesis results on the SERCA pump (Clarke et al., 1989): i.e., the mutated residues in the three transmembrane domains would be involved in the trans-protein Ca^{2+}-channel. However, immunohistochemistry controls have shown that two of the mutations which eliminated Ca^{2+} transport, those of Asn 879 and of Gln 971, also caused retention of the mutated PMCA pumps in the ER (Figure 9). Thus, even if these two residues may be component of the trans-protein Ca^{2+} path, their mutation evidently had structural effects sufficient to alter the plasma membrane targeting of the pump. Evidently, point-mutation experiments of this type, if not accompanied by appropriate controls of non specific (structural) effects, may lead to erroneous interpretations. In conclusion then, the path of Ca^{2+} across the PMCA pump molecule is still unidentified.

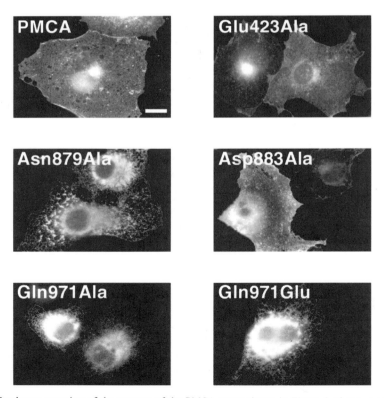

Figure 9. Membrane targeting of the mutants of the PMCA pump shown in Figure 8 after expression in Cos-7 cells. (From Guerini et al., 1996).

5. SUMMARY

The Ca^{2+} pump of the plasma membrane (PMCA) is a high Ca^{2+} affinity enzyme that is regulated by a number of agents. The most important is calmodulin (CaM) which binds to the C-terminal portion of the pump, removing it from an auto-inhibitory site next to the active site. The CaM-binding domain is preceded by an acidic sequence which contains a masked signal for endoplasmic reticulum (ER) retention. A stronger signal for ER retention in the first 45 residues of the sarcoplasmic reticulum (SERCA) pump has been revealed by experiments on chimeras of the two pumps. Four gene products of the PMCA pump are known: two of them (isoforms one and four) are ubiquitously expressed, two are specific for neurons (isoforms two and three) and are induced by the activation of the cells by depolarization. Mutagenesis work has identified 4 residues in transmembrane domains which may be components of the trans-protein Ca^{2+} path. However, the mutation of two of these residues alters the membrane targeting of the pump.

6. REFERENCES

Balàs, R., Gallo R., and Kingsbury, A. (1988). Effect of depolarization on the maturation of cerebellar granule cells in culture. Dev. Brain Res. 40, 269–276.

Brodin, P., Falchetto, R., Vorherr, T., and Carafoli, E. (1992) Identification of two domains which mediate the binding of activating phospholipids to the plasma membrane Ca^{2+} pump. Eur. J. Biochem. 204: 939–946.

Carafoli, E., (1991a) The calcium pump of the plasma membrane. Physiol Rev. 71: 129–153.

Carafoli, E., (1991b) The calcium pump of the plasma membrane. J. Biol. Chem. 267: 2115–2118.

Carafoli, E., (1994) Plasma membrane calcium ATPase: 15 years of work on the purified enzyme. FASEB J. 8: 993–1007.

Carafoli, E., and Guerini, D. (1993). Molecular and cellular biology of plasma membrane Ca^{2+} ATPase. Trends in Cardiovasc. Med. 3: 177–184.

Carafoli, E., Garcia-Martin, E., and Guerini, D., (1996) The plasma membrane calcium pump: recent developments and future perspectives. Experientia 52: 1091–1100.

Caroni, P., and Carafoli, E. (1981). Regulation of Ca^{2+} pumping ATPase of heart sarcolemma by a phosphorylation/dephosphorylation process. J. Biol. Chem. 256: 9371–9373..

Clarke, D.M., Loo, T.W., Inesi, G., and Mac Lennan, D.H., (1989). Location of the affinitiy Ca^{2+}-binding sites within the predicted transmembrane domain of the sarcoplasmic reticulum Ca^{2+}-ATPase. Nature 239: 476–478.

Enyedi, A., Sarkadi, B., Szasz, I., Bot, B., and Gardos, G. (1980). Molecular properties of the red cell calcium pump. II. Effects of proteolysis, proteolytic digestion and drugs on the calcium-induced phosphorylation by ATP in inside / out red cell membrane vesicles. Cell Calcium 1: 299–310.

Falchetto, R., Vorherr, T., Brunner, J., and Carafoli, E. (1991). The plasma membrane Ca^{2+} pump contains a site that interacts with its calmodulin-binding domain. J. Biol. Chem. 266: 2930–2936.

Falchetto, R., Vorherr, T., and Carafoli, E. (1992). The calmodulin-binding site of the plasma membrane Ca^{2+} pump interacts with the transduction domain of the enzyme. Protein. Sci. 1: 1613–1621.

Foletti, D., Guerini, D., and Carafoli, E. (1995). Subcellular targeting of the endoplasmic reticulum and plasma membrane Ca^{2+} pumps: a study using recombinant chimeras. FASEB J. 9: 670–680.

Furukawa, K.I., Tawada, Y., and Shigekawa, M. (1989). Protein kinase C activation stimulates plasma membrane Ca^{2+} pump in cultured vascular smooth muscle cells. J. Biol. Chem. 264: 4844–4849.

Gallo, V., Kingsbury, A., Balàs, R., and Jørgensen, O.S. (1987). The role of depolarization in the survival and differentiation of cerebellar granule cells in culture. J. Neurosci. 7, 2203–2213.

Gopinath, and Vincenzi, F. (1977). Phosphodiesterase protein activator mimics red blood cell cytoplasmatic activator of the $(Ca^{2+} + Mg^{2+})$ ATPase. Biochem. Biophys. Res. Commun. 77:1203–1209.

Guerini, D., Foletti, D., Vellani, F., and Carafoli, E. (1996) Mutation of conserved residues in transmembrane domains 4, 6, and 8 causes loss of Ca^{2+} transport by the plasma membrane Ca2+ pump. Biochemistry 35: 3290–3296.

Heim, R., Iwata, T., Zvaritch, E., Adamo, H.P., Rütishauser, B., Strehler, E.E., Guerini, D., and Carafoli, E. (1992). Expression, purification and properties of the plasma membrane Ca^{2+} pump and of its N-terminally truncated 105-kDa fragment. J. Biol. Chem. 267: 24476–24484.

Hilfiker, H., Guerini, D., and Carafoli, E. (1994). Cloning and expression of isoform 2 of the human plasma membrane Ca^{2+} ATPase. J. Biol Chem. 268: 19717–19725.

Hofmann, F., James, P., Vorherr, T., and Carafoli, E. (1993) The C- terminal domain of the plasma membrane Ca2+ pump contains 3 high affinity Ca^{2+} binding sites. J. Biol. Chem. 268: 10252–10259.

James, P., Vorherr, T., Krebs, J., Morelli, A., Castello, G., Mc Cormick, D.J., De Flora, A., and Carafoli, E. (1989). Modulation of erythrocyte Ca^{2+} ATPase by selective calpain cleavage of the calmodulin binding domain. J. Biol. Chem. 264: 8289–8296.

Jarrett, H.W., and Penniston J.T. (1977). Partial purification of the (Ca^{2+} + Mg^{2+} -ATPase activator from human erythrocytes: its similarity to the activator of 3'-5' cyclic nucleotide phosphodiesterase. Biochem. Biophys. Res. Commun. 77: 1210–1216.

Keeton, T. P., Burk, S.E., and Shull, G.E (1993). Alternative splicing of exons encoding the calmodulin-binding domains and C-termini of plasma membrane Ca^{2+}-ATPase isoforms 1, 2, 3 and 4. J.Biol.Chem. 268, 2740–2748.

Kosk-Kosicka, D., and Bzdega, T. (1988). Activation of the erythrocyte Ca^{2+}-ATPase either by self-association or interaction with calmodulin. J. Biol. Chem. 263: 22–27.

Machamer, C.E. (1993) Targeting and retention of Golgi membrane proteins. Curr. Opin. Cell Biol. 5:606–617.

Monteith, G.R., and Roufogalis, B.D. (1995) The plasma membrane calcium pump - a physiological perspective on its regulation. Cell Calcium 18: 459–476.

Niggli, V., Adunyah, E.S., and Carafoli, E. (1981a). Acidic phospholipids, unsaturated fatty acids, and limited proteolysis mimic the effect of calmodulin on the purified erythrocyte Ca^{2+}-ATPase. J. Biol. Chem. 256:8588–8592.

Niggli, V., Adunyah, E.S., Penniston, J.T., and Carafoli, E. (1981b). Purified (Ca^{2+} + Mg^{2+})-ATPase of the erythrocyte membrane: reconstitution and effect of calmodulin and phospholipids. J. Biol. Chem. 256: 395–401.

Niggli, V., Adunyah, E.S., Penniston, J.T., and Carafoli, E. (1981c). Purification of the (Ca^{2+} + Mg^{2+})-ATPase from human erythrocyte membranes using a calmodulin affinity column. J. Biol. Chem. 254: 9955–9958.

Pedersen, P.L., and Carafoli, E. (1987a). Ion motive ATPases I: ubiquity, properties, and significance to cell function. Trends Biochem. Sci. 12:146–150.

Pedersen, P.L., and Carafoli, E. (1987b). Ion motive ATPases II: energy-coupling and work output. Trends Biochem. Sci. 12: 186–189.

Penniston, J.T., and Enyedi, A. (1994) Plasma membrane Ca2+ pump: recent developments. Cell Physiol. and Biochem. 4: 148–159.

Rega, A.F., and Garrahan, P.J. (1986) The Ca^{2+} pump of plasma membranes, CRC Press. Inc., Boca Raton, FL, USA.

Ronner, P., Gazzotti, P., and Carafoli, E. (1977). A lipid requirement for the (Ca^{2+} Mg^{2+})-activated ATPase of erythrocyte membranes. Arch. Biochem. Biophys. 179: 578–583.

Schatzmann, H.J. (1982). The calcium pump of erythrocytes and other animal cells. in Membrane Transport of Calcium, edited by E. Carafoli. Academic Press, London: 41–108.

Schatzmann, H.J. (1996). ATP-dependent Ca^{++} extrusion from human red cells. Experientia, 22:364–368.

Stauffer, T. P., Hilfiker, H., Carafoli, E., and Strehler, E.E. (1993). Quantitative analysis of alternative splicing options for human plasma membrane calcium pump genes. J. Biol. Chem. 268:25993–26003.

Stauffer, T., Guerini, D., and Carafoli E. (1995). Tissue distribution of the four gene products of the plasma membrane Ca^{2+} pump. J.Biol.Chem. 270, 12184–12190.

Wright, L.C., Chen, S., and Roufogalis, B.D. (1993). Regulation of the activity and phosphorylation of the plasma membrane Ca^{2+}-ATPase by protein kinase C in intact human erythrocytes. Arch. Biochem. Biophys. 306, 277–284.

Zvaritch, E., Vellani F., Guerini, D., and Carafoli, E. (1995). A signal for endoplasmic reticulum retention located at the carboxyl terminus of the plasma membrane Ca^{2+}-ATPase isoform 4Cl. J.Biol. Chem. 270:2679–2688.

8

THE CONCEPT OF PHOSPHATE COMPOUNDS OF HIGH AND LOW ENERGY

Leopoldo de Meis

Instituto de Ciências Biomedicas
Departamento de Bioquímica Medica
Universidade Federal do Rio de Janeiro
Rio de Janeiro, CEP 21.941-590, Brazil

1. INTRODUCTION

Life requires that different forms of energy are continuously interconverted in the cell. For this purpose phosphate compounds are used as the common currency of energy exchange and adenosine triphosphate (ATP) is the principal carrier of energy in the living cell. The hydrolysis of ATP is usually coupled with work. This can be mechanical, as observed in muscle contraction, chemical, as for the synthesis of molecules or osmotic when a gradient is formed across a membrane. It has been shown that several processes of energy conversion are reversible. Thus, several membrane bond ATPase hydrolyze ATP in order to build up an ionic gradient across a membrane and in the reverse process, the energy derived from the gradient can be used to synthesize ATP from ADP and P_i.

We still do not know how energy flows from ATP to the enzyme during the process of energy interconversion. Until recently it was thought that the energy of hydrolysis of ATP and other "high energy" phosphate compounds were the same regardless of whether they were in solution in the cytosol or bound to the enzyme surface and that energy would become available to the enzyme only *after* the phosphate compound had been hydrolyzed. This view was based in calorimetric measurements performed at the beginning of this century which led to the conclusion that energy would only be released to the medium at the moment of cleavage of the phosphate bond (Lipmann, 1941). In this view the sequence of events in the process of energy transduction was thought to be: i) The enzyme binds ATP; ii) ATP is hydrolyzed and energy is released at the catalytic site at the precise moment of cleavage of the phosphate bond; iii) The energy is immediately absorbed by the enzyme; iv) The enzyme uses the energy absorbed to perform work.

For the synthesis of ATP from ADP and P_i, the sequence of events would be the same, but in the reverse order: i) The enzyme would bind ADP and P_i; ii) energy would be released at the catalytic site of the enzyme and used to synthesize ATP; iii) once formed, the ATP molecule would easily dissociate from the enzyme and diffuse into the cytosol

Calcium and Cellular Metabolism: Transport and Regulation, edited by Sotelo and Benech.
Plenum Press, New York, 1997

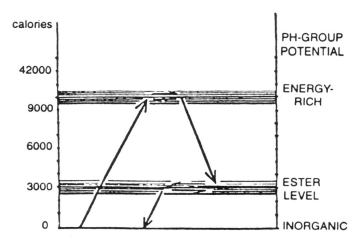

Figure 1. Lipmann's schematic representation for the energy level of different chemical bonds (Lipmann, 1941).

without any further need of energy. From these sequences, it was inferred that work would be performed in a part of the enzyme molecule where energy is released, i.e. in the immediate vicinity of the catalytic site.

During the past three decades the catalytic cycle of different enzymes has been elucidated. These studies revealed that the energy of hydrolysis of different phosphate compounds varies greatly depending on whether they are in solution or bound to enzyme (Table 1).

Reactions that were thought to be practically irreversible in aqueous solution, such as the phosphorylation of glucose by ATP, occur spontaneously when the reactants are bound to the enzyme (Table 1). For enzymes involved in energy transduction, such us the Ca^{2+}-ATPase found in the sarcoplasmic reticulum of skeletal muscle, the energy becomes available for the enzyme to perform work *before* cleavage of the phosphate compound. During the catalytic cycle there is a large decrease of the equilibrium constant for the hydrolysis (K_{eq}) of the phosphate compound bound to the enzyme. In several enzymes stud-

Table 1. Variability of the energy of hydrolysis of phosphate compounds during the catalytic cycle of energy transducing enzymes

Enzyme	Reaction	Solution or enzyme bound, before work		Enzyme bound, after work	
		K_{eq} (M)	G^o kcal/mol	K_{eq} (M)	G^o kcal/mol
Ca^{2+}-ATPase, Na$^+$/K$^+$-ATPase	Aspartyl phosphate hydrolysis	10^6	−8.4	1.0	0
F$_1$-ATPase, Myosin	ATP hydrolysis	10^6	−8.4	1.0	0
Inorganic Pyrophosphatase	PP$_i$ hydrolysis	10^4	−5.6	4.5	−0.9
Hexokinase	ATP + gluc → gluc-P + ADP	2×10^3	−4.6	1.0	0

For details, see reviews in de Meis (1989) and de Meis (1993). The relationship between the standard free energy of hydrolysis (ΔG^o) and the equilibirtum constant of the reaction (K_{eq}) is: $\Delta G^o = -RT \ln K_{eq}$ where R is the gas constant (1.981) and T is the absolute temperature.

ied, work is coupled with this transition of Keq and not with the cleavage of the compound (Table 1). Hydrolysis seems necessary only to permit the dissociation of the nucleotide from the enzyme and not to provide energy to the system. In this new view, the sequence of events for the hydrolysis of ATP is: i) The enzyme binds ATP or other phosphate compounds; (ii) the enzyme performs work without the phosphate compound being hydrolyzed. This is accompanied by a decrease in the energy level of the phosphate compound (iii) the phosphate compound is hydrolyzed in a process which involves relatively small energy change; (iv) the products of hydrolysis dissociate from the enzyme. In the reverse process, phosphate compounds such us ATP and acyl phosphate residue are synthesized on the catalytic site of the enzyme without the need of energy. Energy is then needed for the conversion of the phosphate compound from "low energy" into "high energy". This transition usually occurs on the enzyme surface, before the phosphate compound synthesized being released into the cytosol. There is now a fair amount of information available on the structure of several energy transducing enzymes. These data indicate that work is performed in a region of the tertiary structure of the protein far distant from the protein region were the catalytic site is located and that conformational changes of the protein synchronize the sequence of events occurring in these two regions of the protein. The events responsible for the change in the energy level of phosphate compound at the catalytic site are not clearly understood at present. There is experimental evidence which suggests that one important factor is change in water activity in the environment of the catalytic site (de Meis, 1993. 1989).

The difference between the two views is related to the contribution of enthalpy (ΔH°) and entropy (ΔS°) to the free energy of hydrolysis of ATP according to the equation:

$$\Delta G^\circ = \Delta H^\circ - T\,\Delta S^\circ$$

In the early view, the contribution of entropy was thought to be minimal (Lipmann, 1941), thus free energy and enthalpy would be practically the same. Possible interactions of the phosphate compound with solvent and physiological ions could not be evaluated in calorimetric measurements and thus, were not taken into account. These interactions play an important role in determining the entropy of phosphate compounds hydrolysis. We know now that the K_{eq} for the hydrolysis of PP$_i$ and ATP varies greatly depending on the water activity, pH and divalent cation concentration of the medium. These changes are related mainly to changes of the entropy of the reaction. On the other hand, water activity, pH and divalent cation have practically no effect on the K_{eq} of phosphoesters such us glucose 6-phosphate and phosphoserine (de Meis, 1993, 1989). These measurements and the finding that enzymes may perform work *before* cleavage of ATP indicate that the contribution of entropic energy ($T\,\Delta S^\circ$) may surpass the contribution of enthalpic energy.

In the subsequent sections, the events that led to this new concept will be described in an historical sequence.

2. CELL RESPIRATION AND THE FIRST LAW OF THERMODYNAMICS

In order to understand the evolution of the concept of "high-energy", one must go back to the work of Antoine Lavoisier at the end of the 18th century. Lavoisier was a meticulous accountant in charge of collecting the impost due to the King of France. It was his responsibility to see that the sum of the money collected matched the sum of tax-money due to the realm. Science was the major hobby of Lavoisier, and his activity as account-ant, which earned his living, played a major role in his approach to science (Conant). Be-fore Lavoisier studies, the decrease of solid mass observed during the combustion of organic matter was attributed to the release of "phlogiston", an entity that the alchemists of the middle age believed to be related to the soul of the matter. In the 18th century it was thought that phlogiston was released during both the combustion of organic materials and during the respiration of animals. Lavoisier proved that both process were chemical reac-tions. He weighted the gas and solid matter contained in a closed system before and after combustion and found that the total mass of the system did not vary during combustion. Thus, the hypothetical phlogiston should have a mass and was related to the gases found in the air before and after combustion. During the course of his experiments, Lavoisier discovered a new gas, the oxygen, and demonstrated that in both combustion and respira-tion the oxygen of the atmosphere was consumed with production of carbon dioxide (CO_2). The decrease of mass observed during the combustion of organic matter plus the mass of the oxygen consumed could be accounted by the sum of the masses of water and CO_2 produced. The central feature of the chemical revolution initiated by Lavoisier was the overthrow of the "phlogiston" theory and its replacement by a theory based on the role of oxygen and the new concept known as "conservation of matter" which explains that in a chemical reaction the total mass of all reacting substances must be identically equal to the total mass of the product substances (Conant).

Another important discovery was that the heat released by living animals was de-rived from the oxidative reactions involved in respiration. This conclusion was reached comparing respiration with combustion. In 1780 Lavoisier and Pierre Simon Laplace measured the amount of heat released when equal amounts of CO_2 were produced in the combustion of charcoal and in the respiration of guinea pig. For this experiment, an ice calorimeter was devised by Laplace. They found that the amounts of ice melted in the two processes were practically the same after burning charcoal to yield an amount of CO_2 equal to that exhaled by the animal during respiration. From these experiments, they con-cluded "*Respiration is therefore a combustion, very slow it is true, but otherwise perfectly similar to that of charcoal*". The complete work on animal respiration was published in 1793, a period of despair and cruelty in the French Revolution. Unfortunately, Lavoisier was murdered in the guillotine on May 8, 1794.

The relationship between work and heat was not envisaged by Lavoisier. At that time it was not known that different forms of energy could be interconverted. In fact, at that time energy was starting to be regarded as a physical measurable entity. Before that, energy was referred to as "natural force". The interconversion of energy was first ob-served in a mechanical system by Count B. Rumford. He studied the rise of temperature which accompanied the boring of a cannon and in 1798 concluded that the mechanical work involved in the boring was responsible for the heat produced. The subject did not at-tract any great interest until 1842 when Julius Robert Mayer enunciated the principle of the interconvertibility of energy. Mayer studied medicine at the Tubingen University. He took his degree in 1832 after presenting a dissertation on the effect of santonin on worms in children. In 1840 Mayer signed as ship's doctor and sailed from Rotterdam for a long voyage to Java. According to Mayer's own story, his interest on the relation between heat and work began abruptly on the dock at Surabaya, when several of the sailors needed to be bled (Cohen). The venous blood was such a bright red that at first he thought he had opened an artery. The venous blood of the sailors remained a bright red until they had ac-climated themselves to the tropics. From the work of Lavoisier it was known that the ani-mal heat was generated by a combustion process. Mayer correlated the difference of color between the arterial and venous blood with oxidative reactions and heat production in the blood. He reasoned that in order to maintain the body temperature there should be a relationship between the amount of heat produced, the amount of heat lost and the temperature of the environment. After this initial observation, Mayer started to analyze the values of heat absorbed by expanding gases available in the bibliography and used them to compute the amount of heat equivalent to a given amount of work. From these studies, Mayer stated the principle of the interconvertibility of energy and the conservation of energy. The relationship between the color of the sailors venous blood and the equivalence between heat and work is not readily apparent. Probably, intuition played an important role on Mayer discovery. The first paper he wrote was not accepted for publication. It seems that in this first report, Mayer was confused about the distinctions between the concepts of force, work and energy. His second paper was accepted by Liebing in 1842 for publication in the Annalen der Chemie und Pharmazie. In 1840 J. P. Joules commenced his classical experiments on the relationship between work expended and heat produced. However, the paper of Mayer appeared before any of Joule's results were published. The work of Mayer and of Joules permitted to deduce the conversion factor between mechanical work (w) and heat (q). This is called the mechanical equivalent of heat J:

$$J = w/q$$

In modern units J is usually given as joules per calorie. The combined work of Lavoisier and Mayer permitted them to deduce the first law of thermodynamics, which states that "In Nature nothing is created nor destroyed. All is transformed". Lavoisier, demonstrated the conservation of the matter and that energy is released in biological reactions while Mayer demonstrated the conservation and transformation of energy. It is frequently stated that Lavoisier viewed a living system as a furnace to produce heat while Mayer saw it as a heat engine.

During the next 100 years it became apparent that the energy needed to sustain life should be derived from the cleavage of molecules in the tissues. In order to measure this energy, different substances were burned in a calorimeter and the heat released measured as first reported by Lavoisier and Laplace. At the beginning of this century, long tables with the values of heat of combustion of many different substances were available. From these values, attempts were made to calculate the heat released after the cleavage of the different bonds which unite the atoms of a molecule. For instance, it was estimated that for every electron between C and C, and between C and H 26,050 calories were generated during combustion and that 19,500 calories could be derived from a C:O cleavage (Lipmann, 1941).

At the time of the discovery of the ATP molecule, in the twenties, muscle contraction was the physiological model used to study the process of energy interconversion. In these experiments, calorimetric measurements were correlated with mechanical work. From these studies a paradigm emerged according to which energy would be released when a covalent bound between two atoms of a molecule was cleaved. When the cleavage was catalyzed by enzymes able to transduce energy, as in muscle contraction, then a part of the energy released would be absorbed by the protein and used to perform work and another part could be measured as heat. If the enzyme was not be fit to perform work, then all the energy released would be dissipated as heat. These notions set the frame for the concept of energy-rich bond that followed the identification of the different phosphate compounds of the cell. Though useful at an early stage of bioenergetics, the calorimetric measurement proved later on to be misleading, and the long list of heat of combustion of substances prepared at the beginning of the century were thus found to have little relevance for biological case.

3. THE ATP MOLECULE AND ITS USE IN THE CELL

The history of ATP began in 1847 with the studies of Justus Liebig on the content of meat extracts (Cohen; Moore, 1962). Liebig was a precursor of biotechnology. At that time, meat storage and distribution on a large scale was practically impossible. Liebig prepared a meat extract which could be stored and commercialized. This granted him fame. The Emperor of France Napoleon III, invited Liebig to discuss his studies over dinner and the Emperor of Brazil, Pedro the Second, visited him in Munich and offered a large sum to start a meat industrial process in Brazil (Schlenk, 1987). Liebig become also famous for his bitter criticism of Luis Pasteur elegant experiments on fermentation. Liebig crystallized a barium salt of inosinic acid from a meat extract. The analytical data for carbon, hydrogen and nitrogen permitted the characterization of inosinic acid but Liebig did not detected phosphorus in his measurements. The attachment of phosphorus to the adenosine molecule was first described in 1927 by Embden and Zimmermann who discovered adenosine 5'-phosphoric acid (AMP). ATP was finally isolated as a silver salt in the following year by Fiske and Subbarow (Fiske & Subbarow, 1927; Schlenk, 1987).

The discovery of the physiologic role of ATP is intimately associated with that of creatine phosphate, a compound used to regenerate ATP in the cell:

$$ADP + creatine\ phosphate \rightarrow ATP + creatine$$

In fact creatine phosphate, originally called phosphagen, was discovered by Fiske and Subbarow before ATP (1925) and during several years, most of the effects of ATP were attributed to creatine phosphate. Studying the energetic of muscle contraction both the Eggletons (1927) and Fiske and Subbarow (1927, 1929) found that creatine phosphate was largely decomposed during a long series of contractions and was rapidly reconstituted during recovery of the muscle. At the same time Meyerhof and Suranyi (Lipmann, 1941) found that large amounts of heat were released by enzymatic decomposition of creatine phosphate. In accordance with the paradigm prevailing at the time, these observation naturally lead to the conclusion that energy would be released during cleavage of the phosphate bound of creatine phosphate. For the resynthesis measured during muscle recovery, energy should then be provided by the cleavage of other molecules. Accordingly, Nachmansohn observed that the creatine phosphate cleaved during muscle contraction could be resynthesized at the expense of glycolysis and Lundsgaard (Lipmann, 1941; Lundsgaard, 1932) found that through the breakdown of one-half mole of glucose to lactic acid approximately two moles of creatine phosphate were reformed.

Evidence that creatine phosphate was not directly used for muscle contraction were first reported by Lohmann (1929, 1934). He observed that muscle homogenates were able to catalyze the hydrolysis of creatine phosphate just as the intact tissue does. If the extract was however dialyzed, then the proteins were no longer able to cleave creatine phosphate indicating that there was not a specific enzyme to catalyze its cleavage. Hydrolysis of creatine phosphate was restored if ADP was added together with creatine phosphate to the dialyzed extract. From this observation is was concluded that the true substrate of muscle contraction was ATP and that creatine phosphate was used to regenerate the ATP cleaved. This was confirmed in intact muscle cell with the use of 1-fluoro-2:4-dinitrobenzene (FDNB), a substance which in addition to stop glycolysis also inhibits the transfer of phosphate from creatine phosphate to ADP. Muscles treated with FDNB contract normally and their contraction is accompanied by the breakdown of ATP, the content of creatine phosphate of the muscle remaining unchanged. Shortly after, different laboratories demonstrated that the main product of glycolysis and other catabolic routes is ATP and not creatine phosphate. This led to the conclusion that ATP is in fact the major immediate donor of free energy in biological systems. During catabolism ADP is phosphorylated to ATP and during anabolism ATP is hydrolyzed. Creatine phosphate is a storage form of free energy found in larger amounts in skeletal and cardiac muscle of vertebrates. In its place, most invertebrates contain phosphoarginine, which also serves to regenerate ATP from ADP.

4. THE CONCEPT OF ENERGY-RICH PHOSPHATE BOND

The concept of "energy-rich" and "energy-poor" phosphate compounds was formalized by Lipmann in a review published in 1941 that has become a classical in the bibliography of bioenergetics (Lipmann, 1941). In this review it was formally stated that entropy would play a minor role in determining the thermodynamic parameters of a metabolic reaction. Thus, in the cell the value of free energy would be practically the same as that of enthalpy. The amount of energy which could be derived from the hydrolysis of a phosphate compound would be determined solely by the chemical nature of the bond which links the phosphate residue to the rest of the molecule. The possibility that some energy might be derived from the interaction of reactant and product with the environment (solvent, cations etc.) were not taken into consideration because at that time, the energy of hydrolysis of most phosphate compounds was estimated using calorimetric measurements. The N_P bond of creatine phosphate and the phosphanhydride linkages P-O_P, carboxyl_P and enol_P were identified as energy-rich phosphate compounds, thus having a K$_{eq}$ for the hidrolysis with a high value, ranging from 10^6 to 10^9 M ($\Delta G°$ -8 to -12 kcal/mol). Phosphoesters such as glucose 6-phosphate and glycerol phosphate were refereed to as "energy-poor" phosphate compounds. The K$_{eq}$ for the hydrolysis of phosphoester varies between 10 and 100 and $\Delta G°$ between -1.5 and -2.5 kcal/mol. The methodology used to determine the energies of hydrolysis of the two groups of phosphate compounds were different. For the energy-poor phosphoester the molar concentrations of reactants and products available after the reaction reached equilibrium were measured and from the value of the K$_{eq}$ for hydrolysis, the $\Delta G°$ of the reaction were calculated. Thus, for the hydrolysis of glycerolphosphate measured by Cori et al (1936):

$$\text{glycerolphosphate} + H_2O \rightarrow \text{glycerol} + P_i$$

$$K_{eq} = [\text{glycerol}] \times [P_i] / [\text{glycerolphosphate}] \text{ and}$$

$$\Delta G° = -RT \ln K_{eq}$$

where R is the gas constant (1.981) and T the absolute temperature in which the reaction was performed. This method could not be used for ATP and creatine phosphate because the amount of the energy-rich phosphate compound remaining in solution after that the reaction reaches equilibrium is very small and impossible to measure with the methods available at that time. Therefore free energies were calculated from heat of combustion and heat release during hydrolysis.

In his review (1941) Lipmann introduced the term "group potential" which strongly influenced the course of the theoretical studies in subsequent years. Linkage designed to transfer groups with loss of energy were called "weak" linkages. This included the "energy-rich" phosphate bounds. Lipmann reasoned that if with cleavage, large amounts of energy could be made free, then the tendency to burst the linkage would be relatively great and the linkage would thus be weak, i.e., the phosphate group would have a small affinity for the rest of the molecule. The energy-rich, weak phosphate bond was designated by a squiggle (_ph or N_P). If little energy would be freed with cleavage, or energy has even to be furnished, then the linkage was called strong (large affinity). According to Lipmann proposal, the amount of energy which could be derived from the hydrolysis of a phosphate compound would be determined solely by the chemical nature of the bond cleaved. The transfer of phosphate from one molecule to another would be determined by the energy of hydrolysis of the bond involved.

The γ-phosphate of ATP could be transferred to a molecule of glucose forming glucose 6-phosphate, a compound with a lower energy of hydrolysis than ATP, but the reverse reaction could not occurs without an extra input of energy. Thus, the ATP hydrolyzed in the cell could only be regenerated starting from phosphate compounds having the same, or a higher energy of hydrolysis than ATP itself such us creatine phosphate.

5. THEORETICAL APPROACH

From 1941 until 1969 the theoretical studies of high-energy compounds naturally followed the proposal of Lipmann. Thus, it was thought that intramolecular effects such as opposing resonance, electrostatic repulsions and electron distribution along the P-O-P backbone were the dominant factors contributing to the large negative free energies of hydrolysis of high-energy phosphate compounds such as pyrophosphate and ATP (Kalckar, 1941; Boyd & Lipscomb, 1969).

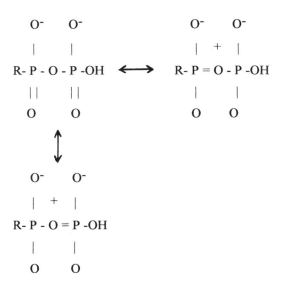

The negative charges on either side of the linkage would repel each other, creating tension within the molecule, and the opposing resonance would generate points of weakness along the P-O-P backbone which tends to stabilize the products relative to reactants. Thus, it would be easy to cleave the molecule and difficult to bring together the products of the hydrolytic reaction. In these formulations water was ignored or regarded as a continuous dielectric for the purpose of calculating repulsion energies.

Table 2. Solvation energies of different ionic forms of inorganic phosphate and pyrophosphate

Molecule	Solvation energy Kcal/mol
$H_2PO_4^-$	76
HPO_4^{2-}	299
PO_4^{3-}	637
$H_3P_2O_7^-$	87
$H_2P_2O_7^{2-}$	134
$HP_2O_7^{3-}$	358
$P_2O_7^{4-}$	584

Values reported by George et al. (1970).

In 1970 George et al. (1970) used a totally different approach. They were the first to propose that interaction of reactants and products with the solvent might play an important role in determining the K_{eq} of a reaction. George et al. reasoned that phosphate compounds interact strongly with water (Table 2). In aqueous solutions, water molecules will organize around the phosphate compound and will both shield the charges of the molecule, thus neutralizing the electrostatic repulsion, and form bridges between different atoms of the molecule, thus reinforcing the weak points generated along the molecule backbone by opposing resonance's. Therefore George and coworkers proposed that the energy of hydrolysis of a phosphate compound would be determined by the differences in solvation energies of reactants and products. Solvation energy is the amount of energy needed to remove the solvent molecules that organize around a substance in solution. Thus, a more solvated molecule would be more stable, i.e., less reactive, than a less solvated molecule and the K_{eq} for hydrolysis have a high value because the products of the reaction are more solvated than the reactant. The solvation energies of orthophosphate and pyrophosphate are shown in Table 2.

In totally aqueous medium and depending on the experimental conditions used, the observed standard energy of hydrolysis ($\Delta G°$) measured for the hydrolysis of pyrophosphate varies between -3 and -6 Kcal/mol (Table 3).

This represents a very small fraction of the total solvation energy of either orthophosphate or pyrophosphate (Table 2). Thus, a small change in the organization of solvent around the molecules of reactants and products might easily lead to a significant change in the thermodynamic parameters of a reaction.

Table 3. Energy of hydrolysis of different phosphate compounds in totally aqueous medium, gas phase and mixtures of water with organic solvents

Compound	$\Delta G°$ Kcal/mol		Organic solvent mixtures
	Water	Gas phase	
ATP	−7.0 to −9.0	—	+0.3
PP$_i$	−3.0 to −6.0	−0.4 to −0.9	−1.0 to +2.0
Aspartyl phosphate	−9.0 to −11.0	+5.0 to +32.0	+0.3 to +2.3
Creatine phosphate	−9.0 to −11.0	+9.0 to +212.0	—
glucose 6-phosphate	−1.5 to −3.0	−1.5 to −2.5	−1.5 to −2.5

The values shown in the table were compiled from Lipmann (1941); de Meis (1984, 1989, 1993); George et al. (1970); Hayes et al. (1978); Ewig & Van Wazer (1988); Flodgaard & Fleron (1974); de Meis et al. (1985); Alberty (1969); and Wolfenden & Williams (1985).

Hayes & Kenyon (1978) calculated the energy of hydrolysis of several phosphate compounds in gas phase and compared these values with those measured in water (Table 3). In aqueous solution, acetyl phosphate and the N-P bounds in both phosphocreatine and phosphoarginine are of a high-energy nature. However, in the gas phase this is no longer true. On the contrary, the large positive ΔH of hydrolysis indicates that when reactants and products are not solvated, acetyl phosphate and phosphocreatine are more stable than the products of their hydrolysis, and according to Lipmann definition, they behave as "energy-poor" phosphate compounds. From these data Hayes et al. (1978) concluded that solvation energies of reactants and products are by far the most important factors in determining the energies of hydrolysis of creatine phosphate, ATP, PP_i and acetyl phosphate. The same conclusion has been reached recently by Ewing and Van Wazer (1988), who calculated the energy of hydrolysis of PP_i in the gas phase. Solvation energy however, does not seems to determine the energy of hydrolysis of all phosphate compounds. According to Hayes et al. calculations (Table 3), the energy of hydrolysis of phosphoesters such us glucose 6-phosphate is the same in water and in gas phase, indicating that the interaction of reactant and products with the solvent does not play a significant role in determining the K_{eq} of hydrolysis of a phosphoesters. A peculiar situation it then becomes apparent. In aqueous solutions an acyl phosphate residue such us that of aspartyl phosphate, and creatine phosphate have a much higher energy of hydrolysis than glucose 6-phosphate, but in gas phase the situation is reversed and glucose 6-phosphate becomes a compound with a higher energy of hydrolysis than either creatine phosphate or aspartyl phosphate.

6. PHOSPHOANHYDRIDE BONDS OF HIGH AND LOW ENERGY

The simplest known "high energy" phosphate compound is pyrophosphate. The only possible product of PP_i hydrolysis is P_i. This greatly facilitates the measurement of the K_{eq} and calculation of the $\Delta G°$ as compared with more complex molecules such as ATP where the ADP produced after hydrolysis can be further cleaved to AMP. Therefore, because of the similarities between the polyphosphate chain of ATP and PP_i, the thermodynamic parameters measured for PP_i are often extrapolated to ATP. Measurements of the K_{eq} (de Meis, 1989; de Meis, 1993; Romero & de Meis, 1989; Flodgaard & Fleron, 1974; de Meis,

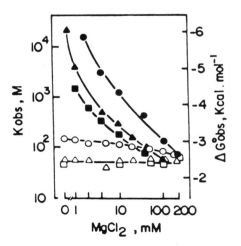

Figure 2. Effect of Mg^{2+} and pH on the energies of hydrolysis of PP_i and phosophoesters. The K_{eq} for hydrolysis of PP_i (filled symbols) and phosphoserine (open symbols) were determined at pH 6.1 (0,0), 7.0 (–,–) and 7.8 (–,–). For details, see de Meis (1993) and Romero & de Meis (1989).

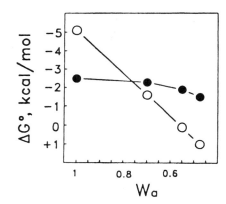

Figure 3. Effect of water activity on the energies of hydrolysis of PP$_i$ and glucose 6-phosphate. (0) PP$_i$; (0) glucose 6-phosphate. For experimental details, see de Meis, (1993); Romero & de Meis (1989); de Meis (1984); and de Meis et al. (1985).

1984; de Meis & Behrens, 1985) revealed that the $\Delta G°$ of PP$_i$ hydrolysis varies greatly depending on the pH and divalent cation concentration in the medium (Fig. 2). A similar variability was found for ATP hydrolysis. By adding the free energy changes of hydrolysis of several different reactions Alberty (1969) calculated that in the pH range of 6.0 to 8.0 and depending on the Mg^{2+} concentrations the $\Delta G°$ of ATP hydrolysis may vary between − 5.0 and −9.9 Kcal/mol. However, pH and Mg^{2+} have practically no effect on the $\Delta G°$ of phosphoesters hydrolysis (Fig.2). Thus, depending on the experimental conditions used, the $\Delta G°$ of PP$_i$ hydrolysis can be either several kcal more negative than that of glucose 6-phosphate or have practically the same value as that of glucose 6-phosphate. A similar pattern was observed when the water activity (w$_a$) of the medium was decreased with the use of organic solvents (Fig. 3). The solvation of molecules in aqueous solution varies depending on the bulk water activity. A decrease of w$_a$ promote a drastic increase of the $\Delta G°$ of both PP$_i$ (Table 3 and Fig. 3) and of ATP (Wolfenden & Williams 1985), but have practically no effect on the energy of hydrolysis of phosphoesters such us glucose 6-phosphate. Thus, in totally aqueous medium (w$_a$ = 1) ATP and PP$_i$ are compounds having a larger energy of hydrolysis than glucose 6-phosphate, but after a small decreased of w$_a$, the situation is inverted and glucose 6-phosphate becomes a phosphate compound having a larger energy of hydrolysis than either ATP or PP$_i$ (de Meis, 1993; Ewig & Van Wazer, 1988; de Meis, 1984; Wolfenden & Williams, 1985).

The values of $\Delta G°$ in Table 3 and Fig.3 are related to the total concentrations of all ionic species of PP$_i$ and P$_i$, including those that are free and those that are in the form of a complex with divalent cations. The change of $\Delta G°$ found at the different pH and Mg^{2+} con-

Table 4. Ionic reactions of PP$_i$ and glucose 6-phosphate hydrolysis

Reaction	$\Delta G°$ Kcal/mol	$\Delta H°$ Kcal/mol	$\Delta S°$ e.u.
$HP_2O_7^{3-} + HOH \rightleftharpoons H_2PO_4^- + HPO_4^{2-}$	−5.2	−8.2	−10.1
$MgP_2O_7^{2-} + HOH \rightleftharpoons MgHPO_4 + HPO_4^{2-}$	−3.3	−8.4	−16.8
$MgP_2O_7^{2-} + HOH \rightleftharpoons 2\,HPO_4 + Mg^{2+}$	−0.6	−11.4	−35.8
$MgHP_2O_7^{2-} + HOH \rightleftharpoons MgHPO_4 + H_2PO_4^-$	−3.6	−8.8	−17.5
$MgHP_2O_7^{2-} + HOH \rightleftharpoons H_2PO_4^- + HPO_4^{2-} + Mg^{2+}$	−0.8	−11.8	−36.1
glucose-P + HOH $\rightleftharpoons$ glucose + $H_2PO_4^-$	−2.4	−1.5	3.1
glucose-P + HOH $\rightleftharpoons$ glucose + HPO_4^2	−2.4	−1.5	3.1
glucose-P + Mg^{2+} + HOH $\rightleftharpoons$ glucose + $MgHPO_4$	−2.4	−1.7	2.5

For details see de Meis (1993); Romero & de Meis (1989); de Meis (1984); and de Meis et al. (1985).

centrations reflects in fact the balance of the different ionic reactions found in each experimental condition. By measuring the K_{eq} of a reaction at different temperatures it is possible to calculate the ΔH and ΔS values of each ionic reaction. Notice in Table 4 that values of ΔH for the different ionic reactions of PP_i hydrolysis varies little. What varies significantly among the different reactions is the ΔS^o value. According to the equation $\Delta G^o = \Delta H^o - T \Delta S^o$ this findings indicate that the large variability of the ΔG^o of PP_i hydrolysis observed when the conditions of the medium are changed is related mostly to enthropic changes of the ionic reactions prevailing in the system and not to a significant change of enthalpy. This is not observed for phosphoesters. The K_{eq} varies little with temperature, and the contribution of entropy to the value of ΔG^o is minimal.

The effects of w_a, pH and magnesium on the ΔG^o of PP_i and glucose 6-phosphate can be easily interpreted according to the solvation energy theory (de Meis, 1989; de Meis, 1993; George & Witonsky, 1970), but difficult to accommodate with the Lipmann proposal (1941). If the energy of the reaction would be determined solely by the nature of phosphate bond cleaved, then the energy of hydrolysis of PP_i should always be higher than that of glucose 6-phosphate regardless of the conditions used. In addition, the effect of pH and magnesium on the energy of hydrolysis of PP_i should be related to the ΔH^o of the different ionic reactions and not to different ΔS^o values as observed (Table 4).

7. TRANSITION BETWEEN BULK SOLUTION AND ENZYME SURFACE

From the data of Fig 2 and Tables 1 and 3 it can be deduced that the thermodynamic parameters of a reaction may vary greatly depending on whether the reactants and products are free in solution or bound to the surface of a protein. Enzyme selectively bind specific forms of the substrate. Physiological solutions contain different ionic forms of ATP, both free and forming complexes with magnesium. However, most energy transducing enzyme selectively bind the complex Mg.ATP. Thus, the ΔG^o of ATP hydrolysis at the catalytic site will be different from that of ATP in solution and similar to that measured in non physiological solutions containing a large excess of magnesium. One of the major differences between bulk solution and the surface of proteins is the activity of the solvent. The water molecules that organize around a protein solution have properties that are different from those of medium bulk water—for example, a lower vapor pressure, a lower mobility and a greatly reduced freezing point. Similar changes in the properties of water are observed in mixtures of organic solvents and water (Cooke & Kuntz, 1974; Saenger, 1987). In addition, the properties of the solvent in a given region of a protein may vary greatly as a consequence of a conformational change. Thus, while the water activity of the bulk solution remains constant, the w_a in the microenvironment of the catalytic site may vary during the catalytic cycle of the enzyme. This has been measured in different transport ATPases (de Meis, 1989).

8. ENTROPIC ENERGY AND TRANSPORT ATPases

In equilibrium thermodynamics, as proposed by Ludwig Boltzmann in 1877, entropy is a measure of disorder. In a closed system, entropy increases as the system flows from the initial state toward equilibrium. For an osmotic gradient, the entropy increases as the concentration of solutes is equalized in the two compartments. To restore the gradient, and thus to decrease once again the entropy, energy must be provided to the system. Transport ATPases operate based in this principle. During active transport these enzymes use the energy derived from ATP to form a gradient across the membrane and in the reverse process the increase of entropy associated with the decrease of the gradient is used to synthesize ATP. This was shown first by Peter Mitchell (1961) who demonstrated that the ATP F$_1$-F$_o$ complex of mitochondria, chloroplasts and bacteria catalyzes the synthesis of ATP from ADP and P$_i$ using the energy of the electrochemical proton gradient derived from electron transport. In these organelles, ATP is first spontaneously formed from ADP and P$_i$ at the catalytic site of the enzyme without the need of an energy input. The ATP thus formed has a low energy of hydrolysis and can not dissociate from the enzyme. The energy derived from the electrochemical proton gradient is needed to both increase the K$_{eq}$ for the hydrolysis of the tightly bound ATP from 1 to a value higher than 10^{-6}M and to permit the dissociation of the ATP synthesized from the enzyme (Boyer & Ross, 1973; Penefsky & Cross, 1991).

The catalytic cycle of transport ATPases is simpler than that of the ATP synthase and is better understood at present. This include the Ca^{2+}-transport ATPase found in the sarcoplasmic reticulum of skeletal muscle (de Meis & Vianna, 1979; Tanford, 1984):

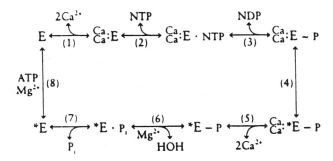

This Ca^{2+}- ATPase can catalyze both the hydrolysis and the synthesis of ATP. The chemical energy derived from the hydrolysis of ATP is used to form a Ca^{2+} gradient across the membrane and, in the reverse process, the energy derived from the osmotic gradient is used to synthesize ATP from ADP and P$_i$. For the hydrolysis of ATP, the catalytic cycle is initiated after that ATP and Ca^{2+} ions bind to two different domains of the enzyme (steps 1 and 2 in the sequence). The catalytic site of the enzyme, which binds selectively the complex Mg.ATP, is located in a hydrophilic region of the protein protruding from the membrane into the cytosol. Two calcium ions bind in a different region of the protein immersed in the hydrophobic moiety of the membrane (Clarke & Loo, 1989). After the binding of Mg, ATP and Ca^{2+}, an aspartyl residue located at the catalytic site is phosphorylated by ATP forming an acylphosphate residue (step 3). The K$_{eq}$ of this reaction is close to one and the K$_{eq}$ for the hydrolysis of the acylphosphate residue formed is higher than 10^6M, therefore similar to that of ATP in solution. After phosphorylation, the protein undergoes a spontaneous conformational change which is felt both at the catalytic site and on the Ca^{2+}-binding domains of the ATPase (step 4). During this event, the calcium ions bound to the protein are translocated across the membrane. This is followed by a large decrease of the affinity of the protein for calcium which permits the two calcium translocated to dissociates from the protein on the other side of the membrane (step 5). Simultaneously with the translocation, at the catalytic site there is a large decrease of the K$_{eq}$ for the hydrolysis of the acylphosphate residue. This is accompanied by a decrease of water activity in the microenvironment of the catalytic site (de Meis, 1989). The calculated value for the energy of hydrolysis of an acylphosphate residue in gas phase is similar to that measured for the acyl-phosphate residue after the conformational change of the protein (Tables 1 and 3), indicat-ing that the decrease of the energy was promoted by a change of solvation of the reactant and products at the catalytic site. The final events are the hydrolysis of the acylphosphate residue, dissociation of P_i from the enzyme (steps 6 and 7) and the spontaneous return of the enzyme to the original conformation (step 8). There is little or no energy exchange in these final steps and the enzyme is ready to initiate a new cycle.

The sequence of events for the synthesis of ATP is the same as that for the hydroly-sis but in the reverse order. This is observed when there is a large difference of Ca^{2+} con-centration on the two sides of the membrane, less than 10^{-6} M in the cytosolic side and higher than 10^{-3} on the other side of the membrane (Barlogie & Hasselbach, 1971; Maki-nose & Hasselbach, 1971). The enzyme is initially phosphorylated by P_i forming a low en-ergy acylphosphate residue (steps 7 and 6). There is no need of energy for this step, the K_{eq} for this reaction being close to unity (Masuda & de Meis, 1973). Ca^{2+} must then bind on the non-cytosolic side of the membrane (step 5). The energy derived from this binding permits the conformational change (step 4) which leads to an increase of the water activity in the catalytic site and conversion of the phosphoenzyme from low into high energy (de Meis & Carvalho, 1974). This acylphosphate residue transfers its phosphate fo ADP form-ing ATP which dissociates from the enzyme (steps 3 and 2). Simultaneously the calcium is translocated across the membrane in a process which involves a large increase in the affin-ity of the protein for the calcium bound (step 4). A very low Ca^{2+} concentration in the cy-tosolic side of the membrane is then needed to permit the dissociation from the ATPase of the tightly bound calcium (step 1) and initiation of a new catalytic cycle (step 8). In this system, the transduction of energy occurs during the conformational change which change the energy level of the phosphoenzyme (step 4). During the hydrolysis of ATP the confor-mational change which permits the conversion of the phosphoenzyme from high into low energy occurs spontaneously, but in the reverse direction, the energy derived from the binding of Ca^{2+} to the low affinity sites of the ATPase (step 5) is needed to permit the con-version of the phosphoenzyme from low into high energy. Notice however that in both cases the process of energy transduction is not associated with the cleavage or formation of the acylphosphate bond. During the active transport calcium is translocated across the membrane *before* the cleavage of the bond and during synthesis, energy is provided by the binding of calcium *after* the formation of the acylphosphate bond. A curious feature of the Ca^{2+} transport ATPase is that the enzyme can synthesize ATP not only when a Ca^{2+} gradi-ent is formed across the membrane but also when either a H^+, temperature or water gradi-ent are formed (de Meis, 1989; de Meis & Martins, 1980; de Meis & Inesi, 1985). As discussed above, the energy of hydrolysis of phosphoanhidrides varies depending on the divalent cation concentration, pH, w_a and temperature. These parameter change the contri-bution of entropic energy ($T \Delta S°$) of the reaction. Therefore, it seems that the Ca^{2+}-ATPase can detect different entropic changes on the two sides of the membrane and used it to syn-thesize ATP from ADP and P_i. The conformational change which permits the conversion of the phosphoenzyme from low into high energy is mediated by the binding of Ca^{2+} to the enzyme. The affinity of the enzyme for Ca^{2+} varies depending on the pH and temperature in contact with each side of the membrane. A 2 pH unit increase of the medium leads to a 100-fold increase of the enzyme affinity for Ca^{2+}. A similar increase is observed when the temperature is decreased from 30° to 0° C. Thus, the ATPase can synthesize ATP when there are equal Ca^{2+} concentrations on the two sides of the membrane but different pH or temperature which alters the Ca^{2+} affinity and thus the Ca^{2+} concentration needed to pro-mote the conversion of the phosphoenzyme from low into high energy. Changes of water activity operates in a different manner (de Meis, 1980; de Meis & Inesi, 1985). The conformation of the enzyme which permits the formation of the low energy phosphoenzyme is only available when there is no binding of Ca^{2+} on the cytosolic side of the membrane. A decrease of w_a in the cytosolic side of the membrane greatly facilitates the partition of P_i from the medium into the catalytic site of the ATPase, thus increasing more than 5 order of magnitude the affinity of the enzyme for P_i. In this condition the ATPase can be phosphorylated by P_i even when there are equal Ca^{2+} concentration on the two sides of the membrane. An hydrophobic-hydrophilic transition is then needed to permit the change of water activity inside the catalytic site and conversion of the phosphoenzyme from low into high energy. Thus, the ATPase can catalyze the synthesis of ATP when the Ca^{2+} concentration is maintained high on the two sides of the membrane and the w_a changes in cycles around the enzyme.

The reaction sequence described for the sarcoplasmic reticulum of skeletal muscle is characteristic of different transport ATPases. These include the $(Na^+ + K^+)$ ATPase and calmodulin regulated Ca^{2+} transport ATPase of plasma membrane, the H^+K^+ ATPase of the gastric mucous and the H^+ ATPase of plants. These enzyme are referred as $E_1 - E_2$ enzymes because all of them undergo a conformational change during the catalytic cycle with formation of an acylphosphate residue having different energies of hydrolysis: E_1 high energy and E_2 low energy (Pedersen & Carafoli, 1987).

9. ATP REGENERATING SYSTEMS OF HIGH AND LOW ENERGY

There are physiological conditions in which the cell may need to increase suddenly the consume of ATP. An example is a rapid and intense muscle contraction. In these conditions the ATP hydrolyzed can be readily regenerated from other phosphate compounds such us creatine phosphate or ADP.

$$ADP + \text{creatine phosphate} \rightarrow ATP + \text{creatine}$$
$$2ADP \rightarrow ATP + AMP$$

These reactions are catalyzed by creatine kinase and adenylate kinase respectively, and permits the rapid recovery of the cytosolic ATP level before the resynthesis through more complex metabolic routes such us glycolysis in the cytosol or the oxidative phosphorylation in mitochondria are activated. These are the classic ATP regenerating system and during several years it was thought that ATP could only be regenerated starting from phosphate compounds having a higher, or at least, the same energy of hydrolysis than ATP itself. This notion is derived from the original proposal of Lipmann in 1941 which assumed that the energy of hydrolysis of a phosphate compound would not vary in the cell. Within this view hexokinase could not catalyze the reaction

$$\text{glucose 6-phosphate} + ADP \rightarrow ATP + \text{glucose} \quad (\Delta G^\circ +4.5 \text{ Kcal/mol})$$

because the enzyme can only bind one molecule of ADP for each glucose 6-phosphate molecule bound and the unfavorable Keq of the reaction ($6x10^{-4}$) would require that for each ADP molecule, more than a thousand molecules of glucose 6-phosphate should be crowd inside the catalytic site of hexokinase to permit the formation of one molecule of ATP. Thus the enzyme would only be able to promote the phosphorylation of glucose from ATP but never the reversal, i.e., synthesis of ATP from glucose 6-phosphate and ADP.

Contrasting with this view, Wilkinson and Rose (1979) discovered that on the surface of hexokinase the K_{eq} of the reaction is one (Table 1), indicating that on the enzyme surface the energy of ATP hydrolysis has a similar value than that of glucose 6-phosphate as shown by the sum of the equations below:

$$\text{glucose 6-phosphate} + \text{HOH} \rightarrow P_i + \text{glucose}$$
$$\underline{\text{ADP} + P_i \rightarrow \text{ATP} + \text{HOH}}$$
$$\text{glucose 6-phosphate} + \text{ADP} \rightarrow P_i + \text{ATP}$$

The finding of Wilkinson and Rose is readily explained by the solvation theory. As shown in Table 3 and Fig. 3, a small decrease of water activity promotes a decrease of the energy of ATP hydrolysis to a level similar or even lower than that of glucose 6-phosphate. This permits the enzyme to promote the transfer reaction but after dissociation from the enzyme, the concentration of ATP which can be accumulated in the solution after equilibrium is very small as compared to that of ADP because in water the difference between the energies of hydrolysis of ATP and glucose 6-phosphate becomes once again large. Several enzymes involved in energy transduction however, possess a very high affinity for ATP. Examples are the transport ATPases, such us the Ca^{2+}-ATPase and the ATP synthase complex of mitochondria and chloroplasts. The K_a for ATP binding at the catalytic site of these enzymes is 10^{-7} and 10^{-12} M, respectively (Boyer & Ross, 1973; Penefsky & Cross, 1991; Srivastava & Bernhard, 1987; de Meis & Monteiro-Lomeli, 1992). The affinity of these two enzymes is sufficiently high to permit the formation of the enzyme-substrate complex even in presence of the very small concentrations of ATP formed from ADP and glucose 6-phosphate. After each catalytic cycle, the ATP hydrolyzed by the transport ATPases is rephosphorylated by glucose 6-phosphate to maintain the equilibrium concentration of ATP. In steady-state conditions, the concentrations of ATP remain constant, and the work performed by the two transport ATPases is coupled to a decrease in the glucose 6-phosphate concentration, a compound that has a smaller energy of hydrolysis than does either ATP or creatine phosphate. Once formed, the energy derived from the gradients can in turn be used to promote either: (1) the synthesis of ATP from ADP and P_i, to a concentration much higher than that possible after the equilibrium of the hexokinase reaction. This was measured with the Ca^{2+} transport ATPase; (2) the uphill electron transfer from succinate to NAD^+ in mitochondria. These reactions have an energy requirement higher than that derived from the simple cleavage of glucose 6-phosphate (Monteiro-Lomeli & de Meis, 1992; de Meis & Monteiro-Lomeli, 1992).

10. CONCLUSIONS

Instead of the classical definition of ATP as being a high energy phosphate compound, we think that a more accurate definition would be that ATP and other phosphate molecules possessing phosphoanhydride bonds are molecules that permit the use of entropic energy. Contrasting with phosphoanhydride bonds, the entropy for the hydrolysis of phosphoesters such us glucose 6-phosphate and phosphoserine practically does not vary after large variations of water activity, salt concentrations or pH values of the medium. Therefore, the energy of hydrolysis of these compounds will be the same during enzyme catalysis, i.e., in the transition from the aqueous medium to the surface of the enzyme and then back again to the solution. Instead of referring to phosphoesters as "low energy", we may define them as phosphate compounds that do not permit the use of entropic energy.

While having a K$_{eq}$ for the hydrolysis much smaller than that of ATP in aqueous solution, in the surface of enzyme phosphoester may have an energy of hydrolysis higher than that of ATP. This variability permits the transfer of the phosphate from glucose 6-phosphate to ADP forming ATP in the surface of hexokinase. Therefore, phosphate compounds having quite different energies of hydrolysis in water, such us glucose 6-phosphate and creatine phosphate, can be used by enzymes to regenerate ATP from ADP. The difference between the two systems is however the ATP:ADP ratio found in the solution after equilibrium is reached. With creatine phosphate practically all the nucleotide is in the form of ATP while with glucose 6-phosphate, only a very small fraction is ATP and most of the nucleotide found in solution is ADP. In the particular case of enzymes having a very high affinity for ATP, this does not represent an impediment for the use of glucose 6-phosphate and hexokinase as an ATP regenerating system since they can recognize and use the very low ATP concentration available in the medium. Thus, in steady state conditions the work performed by an enzyme involved in energy transduction such as the Ca^{2+} transport ATPase of skeletal muscle, can be ultimately coupled with the cleavage of glucose 6-phosphate, a phosphate compound that in aqueous solution has and energy of hydrolysis smaller than that of creatine phosphate and ATP.

11. SUMMARY

During the past three decades it was discovered that the energy of hydrolysis of ATP and other phosphate compounds possessing a phosphoanhydride bond such us PP$_i$ and acyl phosphate residue, varies greatly depending on the water activity and ionic composition of the medium in which they are found. The variability is related to changes of the entropy of the hydrolysis reaction. In aqueous solutions these compounds have a high K$_{eq}$ for the hydrolysis but in conditions similar to those found on the surface of proteins they have a low K$_{eq}$ for the hydrolysis, thus behaving as "low energy" phosphate compounds. The work performed by proteins is associate with this transition of the K$_{eq}$ in the enzyme surface, i.e., with the conversion of the phosphate compound from high into low energy, and not with the cleavage of the phosphoanhydride bond at the catalytic site of the enzyme.

In the surface of enzymes phosphoester such us glucose 6-phosphate have an energy of hydrolysis higher than that of ATP. This permits the transfer of the phosphate from glucose 6-phosphate to ADP forming ATP in the surface of hexokinase. Enzyme that have a very high affinity for ATP, as for instance the Ca^{2+}-transport ATPase of the sarcoplasmic reticulum and the F1-Fo complex of mitochondria, can use glucose 6-phosphate and hexokinase as an ATP regeneratins system.

12. REFERENCES

Alberty, R.A. (1969). Standard Gibbs free energy, as a function of pH and pMg for reactions involving adenosine phosphates. J.Biol.Chem. 244:3290–3302.

Barlogie, B., Hasselbach, W., & Makinose, M. (1971). Activation of calcium effux by ADP and inorganic phosphate. FEBS Lett. 12:267–268.

Boyd, D. B. & Lipscomb, W. N. (1969). Electronic Structures for Energy-Rich Phosphates. J.Theor.Biol. 25:403–420.

Boyer, P.D., Ross, R.L., & Momsen, W. (1973). A new concept for coupling in oxidative phosphorylation based on the oxigen exchange reactions. Proc. Natl. Acad. Sci. USA 70:2837–2839.

Clarke, D.M., Loo, T.W., Inesi, G., & MacLennan, D.H. (1989). Location of high affinity Ca^{2+} binding sites within the predicted transmembrane domain of the sarcoplasmic reticulum Ca^{2+} ATPase. Nature 339:476–478.

Cohen, I. Bernard. In Revolution in science, pp 229–236. The Belknap Press of Harvard University Press, Cambridge, Mass.

Cooke, R., & Kuntz, I.D. (1974). The properties of water in biological systems. Annu.Rev.Biophys. Bioeng. 3:95–107.

Cori, G.T., Cori, C.F. & Schmidt, G.(1936). Mechanism of formation of hexosemonophosphate in muscle and isolation of a new phosphate ester. Proc .Soc. Exp. Biol. and Med. 34:702–708.

de Meis, L. (1984). Pyrophosphate of high and low energy: Contribution of pH, Ca^{2+}, Mg^{2+} and water to free energy of hydrolysis. J. Biol. Chem. 259:6090–6097.

de Meis, L. (1989). Role of water in the energy of hydrolysis of phosphate compounds—Energy Transduction in Biological Membranes. Biochem. Biophys. Acta (Reviews) 973:333–344.

de Meis, L. (1993). The concept of energy-rich phosphate compounds: water, transport ATPases and entropic energy Arch. Biochem. Biophys. 306:287–296.

de Meis, L. Grieco, M.A., & Galina, A. (1992). Reversal of oxidative phosphorylation in submitochondrial particles using glucose 6-phosphate and hexokinase as an ATP regenerating system. FEBS Lett. 308:197–201.

de Meis, L., & Carvalho, M.G.C.(1974). Role of the Ca^{2+} concentration gradient in the adenosine 5'triphosphate. Inorganic phosphate exchange catalyzed by sarcoplasmic reticulum. Biochemistry 13:5032–5038.

de Meis, L., & Inesi, G. (1982). ATP synthesis by sarcoplasmic reticulum ATPase following Ca^{2+}, pH temperature and water activity jumps. J. Biol. Chem. 257:1289–1294.

de Meis, L., & Inesi, G. (1985). Enzyme phosphorylation with Pi causes Ca^{2+} dissociation from sarcoplasmic reticulum ATPase. Biochemistry 24:922–925.

de Meis, L., & Tume, R.K. (1977). A new mechanism by which and H^+ concentration gradient drives the synthesis of ATP, pH jump, and ATP synthesis by the Ca^{2+}-dependent ATPase of sarcoplasmic reticulum Biochemistry 16:4455–4463.

de Meis, L., & Vianna, A.L. (1979). Energy interconversion by the Ca^{2+}-transport ATPase of sarcoplasmic reticulum. Annu.Rev.Biochem. 48:275–292.

de Meis, L., Martins, O.B. & Alves, E.W. (1980). Role of water, hydrogen ions, and temperature on the synthesis of adenosine triphosphate by the sarcoplasmic reticulum Adenosine Tryphosphatase in the absence of a calcium ion gradient. Biochemistry 19:4252–4261.

de Meis, L., Monteiro-Lomeli, M., Grieco, M.A., and Galina, A. (1992).The Maxwell Demon in Biological systems—Use of glucose 6-phosphate and hexokinase as an ATP regenerating system by the Ca^{2+}-ATPase of sarcoplasmic reticulum and submitochondrial particles. Ann. N.Y. Acad. Sci. 671:19–31.

de Meis, L.,Behrens, M.I., Petretski, J.H., & Politi, M.J. (1985). Contribution of water to free energy of hydrolysis of pyrophosphate. Biochemistry 24:7783–7789.

Eggleton, P. & Eggleton, G.P. (1927). Inorganic phosphate and labile form of organic phosphate in gastrocnemius of frog. Biochem. J. 21:190–194.

Ewig, C.S. &Van Wazer, J.R. (1988). Ab Initio structures of phosphorus acids and esters. 3. The P-O-P Bridged compounds $H_4P_2O_{2n=1}$ for n=1 to 4. J. Am. Chem. Soc. 110:79–86.

Fiske, C.H. & Subbarow, Y. (1925) The colorometric determination of phosphorus. J. Biol. Chem. 66:375–381.

Fiske, C.H. and Subbarow, Y. (1927). The nature of the "inorganic phosphate" in voluntary muscle. Science 65:401.

Fiske, C.H. & Subbarow, Y. (1929). Phosphorus compounds of muscle and liver. J.Biol.Chem. 81:629–635.

Flodgaard, H., &Fleron, P. (1974). Termodynamic parameters for the hydrolysis of inorganic phosphate at pH 7.4 as a function of $[Mg^{2+}]$, $[K^+]$, and ionic strength determined from equilibrium studies of the reaction. J.Biol.Chem. 249:3465–3474.

George, P., Witonsky, R.J., Trachtman, M., Wu, C., Dorwatr,W., Richman, L., Richman, W., Shuray, F. & Lentz, B.(1970) An enquiry into the importance of solvation effects in phosphate ester and anhydride reactions. Biochim.Biophys.Acta 223:1–15.

Hayes, M.D., Kenyon, L.G., and Kollman, A.P. (1978). Theoretical calculations of the hydrolysis energies of some "high-energy" molecules. J.Am.Chem.Soc. 100:4331–4340.

Hill, T.L. & Morales, M.H.(1951). On 'High Energy Phosphate Bonds' of Biochemical Interest. J.Am.Chem.Soc. 73:1656–1660.

Kalckar,M. (1941). An activator of the hexokinase system. J. Biol.Chem. 137:789–790.

Lipmann, F. (1941) Metabolic Generation and utilization of phosphate bond energy. Advances in Enzymology 1:99–162.

Lohmann,K. (1929). The chemistry of muscle contraction. Naturwiss. 17:624–626.

Lohmann,K. (1934).The chemistry of muscle contraction. Naturwiss. 22:409–411.

Lundsgaard, E. (1932). The significant of the phonomenon "alactacid muscle contractions" for an interpretation of the chemistry of muscle contraction. Danske Hospitalstidennde, 75:84–87.

Makinose, M., & Hasselbach, W. (1971). ATP synthesis by the reverse of the sarcoplasmic reticulum pump. FEBS Lett. 12:271–272.

Masuda, H., & de Meis, L. (1973). Phosphorylation of the sarcoplasmic reticulum membrane by orthophosphate. Inhibition by calcium ions. Biochemistry. 12:4581–4585.

Meyerhof, O. (1930). Conversion of fermentable hexoses with a yeast catalyst. (hexokinase). J.Springer, Berlin, pp. 149–155.

Mitchell, P. (1961). Coupling of phosphorylation to electron and hydrogen transfer by a chemiosmotic type of mechanism. Nature 191:144–148.

Montero-Lomeli, M., & de Meis, L. (1992). Glucose-6-phosphate and hexokinase can be used as an ATP regenerating system by the Ca^{2+} ATPase of Sarcoplasmic Reticulum. J. Biol. Chem. 267:1829–1833.

Moore, Walter J. (1962) Physical Chemistry. Prentice-Hall Inc., Englewood Cliffs, N.J.

The Overthrow of the Phlogiston Theory. Harvard Case Histories in Experimental Sciences, J.B.Conant, ed. Harvard University Press, Cambridge, Mass.

Pedersen, P.L., & Carafoli, E. (1987). II. Energy coupling and work output. Trends Biochem. Sci. 12:145–147.

Pedersen, P.L., & Carafoli, E. (1987). Ion motive ATPases. I. Ubiquiti, properties, and significance to cell function. Trends Biochem. Sci. 12:145.

Penefsky, H.S., & Cross, R.L. (1991). Structure and mechanism of F$_0$F$_1$-type ATP syntases and ATPases. Adv. Enzymol. Relat. Areas Mol. Biol. 64:174–214.

Romero, P., & de Meis, L. (1989). Role of water in the energy of hidrolysis of phospho-anhydride and phosphoesters bonds. J. Biol. Chem. 264:7869–7873.

Saenger, W. (1987). Structure and dynamics of mater surrounding biomolecules. Annu. Rev.Biophys. Biophys. Chem. 16:93–102.

Schlenk, F. (1985) Early research on fermentation—a story of missed opportunities. TIBS 10:252.

Schlenk, F. (1987). The ancestry, birth and adolescence of adenosine triphosphate. TIBS 12:367–368.

Srivastava, D.K., & Bernhard, S.A. (1987). Biophysical chemistry of metabolic reaction sequences in concentrated enzyme solution and in the cell. Annu. Rev. Biophys. Biophys. Chem. 16:175–187.

Tanford, C. (1984). Twenty questions concerning the reaction cycle of the Sarcoplasmic Reticulum calcium pump. Crit. Rev. Biochem. 17:123–151.

Wilkinson, K.D., & Rose, I.A. (1979). Activation of yeast hexokinase PII. Changes in conformation and association. J. Biol. Chem. 254:2125–2129.

Wolfenden, R., & Williams, R. (1985). Solvent water and the biological group-transfer potential of phosphoric and carboxylic anhydrides. J. Am. Chem. Soc. 107:4345–4346.

9

UNDERSTANDING THE ENERGY SOURCE FOR Na$^+$-Ca^{2+} EXCHANGE AFTER DEPHOSPHORYLATION STEPS OF THE Na$^+$-ATPase ACTIVITY OF Na$^+$, K$^+$-ATPase

Luis Beaugé, Marta Campos, and Roberto Pezza

División de Biofísica
Instituto de Investigación Médica "Mercedes y Martín Ferreyra"
Casilla de Correo 389, 5000 Córdoba, Argentina

1. INTRODUCTION

Living cells maintain a steady different ionic compositions between the intra and extracellular compartments. This implies electrochemical gradients for many ionic species which are of large magnitudes; in turn, it requires mechanisms that keep a delicate balance between inward and outward ionic fluxes across the cellular membrane. Dissipating fluxes in favor of an electrochemical gradient are opposed by energy consuming fluxes of the same magnitude. In the case of the so called "pumps" the free energy comes directly from ATP hydrolysis. In the co- and countertransport systems the energy arises from the gradient dissipation of other ionic species (Ullricht, 1979; Tanford, 1983; Aroson, 1985). A paradigmatic case is given by Ca^{2+} ions. With an electronegative cell interior the intracellular Ca^{2+} concentration is about 10^4 times smaller than that at the extracellular fluid. Two mechanisms work in parallel to account for this large electrochemical gradient. The Ca^{2+} pump and the Na$^+$-Ca^{2+} exchange. The first takes the required energy from ATP. The second extrudes Ca^{2+} at the expenses of the free energy stored in the gradient of Na$^+$ across the membrane. Actually, the Na$^+$ gradient is as a generalized energy donor for co- and countertransport of solutes as it is the ATP for ionic pumps. Therefore, the knowledge of the mechanism for the Na$^+$ electrochemical gradient generation is of paramount importance for all non-pumped energy requiring transports mechanism. This chapter deals with a detailed study of some partial reactions of the complex Na$^+$-K$^+$ transport cycle which structure, transport and biochemical identity is the Na$^+$,K$^+$ATPase (Glynn, 1985; Norby & Klodos, 1988; Froehlich & Fendler, 1991).

2. THE Na$^+$, K$^+$-ATPase ENZYME

The Na$^+$,K$^+$-ATPase, also known as Sodium Pump, belongs to the family of E$_1$-E$_2$ enzymes (or P-Type Transport ATPases). Its function is to export Na$^+$ ions from the cell in ex-

Calcium and Cellular Metabolism: Transport and Regulation, edited by Sotelo and Benech.
Plenum Press, New York, 1997

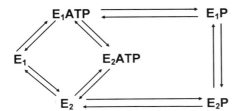

Figure 1. Albers-Post reaction cycle for the Na^+,K^+-ATPase activity. In order to simplify the scheme the bindings of Na^+, K^+, Mg^{2+} and ADP were not included.

change for K^+ ions from the extracellular space. The overall cycle (see Figure 1) occurs at a ratio of 3 Na^+ ions expelled and 2 K^+ ions taken up for each ATP molecules hydrolyzed. This system can perform different total and partial reactions; in some cases there is a biochemical counterpart to the transport event, while in others one of them is missing (biochemical phenomenon not associated to transport or transport without a biochemical counterpart). Actually, a lot of what is known about the Sodium Pump has come from the study of reactions others that the total ($Na^+ + K^+$)-stimulated hydrolysis. Depending on the ionic environment bathing the cell membrane, two main routes of overall ATP hydrolysis can be obtained. In the presence of $Na^+(i)$ and $K^+(o)$, the ($Na^+ + K^+$)-activity associated to the overall cycle of Na^+-K^+ exchange. Intracellular Na^+ is essential for phosphorylation of the E_1 form from ATP in the presence of Mg^{2+} (K_m for ATP below 1 μM); it is believed that the same Na^+ is then expelled to the extracellular medium. External K^+ accelerates the breakdown of a phosphoenzyme already in the E_2 conformation and is subsequently translocated into the cell. In their transit through the membrane both Na^+ and K^+ participate in enzyme conformations designated "occluded" (Beaugé & Glynn, 1979a; Forbush, 1988; Glynn & Karlish, 1990; Beaugé et al., 1990); for Na^+ is a phosphorylated form, likely $MgE_1P(Na_3)$, and for K^+ a dephospho state, known as $E_2(K_2)$. Under normal conditions Na^+ occlusion is very unstable and cannot be detected in native enzyme. On the contrary, the release of K^+ from $E_2(K_2)$ (the $E_2(K_2)$-E_1K_2 transition and the release of K^+ from E_1) is slow (about 0.02–0.04 s^{-1} at 20°C); that rate is increased about 500–1000 fold by ATP acting with low affinity in a non phosphorylating regulatory role. This means that while Na^+ extrusion follows one single path, K^+ uptake has two optional routes depending on ATP regulation; one slow and one fast. Under physiological conditions (Na^+, K^+ and both ATP sites at or near saturation) the rate limiting step is K^+ deocclusion; the overall biochemical and transport reactions show an ATP activation curve with $K_{1/2}$ about the K_m of the regulatory site (100–300 μM). Without external K^+ and with intracellular Na^+ there is what we know as Na^+-ATPase activity; in this case the steps involved in phosphorylation and Na^+ extrusion do not suffer major changes, but the return path, beginning with enzyme dephosphorylation is drastically altered. This may happens in two ways: In the absence of extracellular Na^+, the Na^+-ATPase (or Na^+, 0-ATPase) activity is accompanied by an uncoupled Na^+ efflux (Glynn & Karlish, 1976). When present extracellularly, Na^+ ions act like K^+ stimulating enzyme dephosphorylation and being transported inwardly with in the same 3:2 coupling ratio (Na^+, Na^+-ATPase, Cornelius, 1991). However, little was known about the actual fate of the after dephosphorylation steps under these last two conditions.

3. THE PROPOSED Na^+-ATPase CYCLE

3.1. The Model

Working at 0°C, under K^+-free conditions and in the presence of Na^+ and Mg^{2+}, Post et al. (1975) observed that there was no Pi incorporation into the Na^+, K^+-ATPase. On the

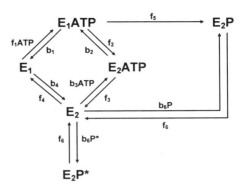

Figure 2. Working model for the Na$^+$-ATPase cycle of the Na$^+$,K$^+$-ATPase. In order to simplify the scheme the E$_1$P intermediate and the bindings of Na$^+$, K$^+$ and Mg^{2+} were not included.

other hand, a sizable Pi-dependent phosphorylation occurred when ATP was added. They suggested that without ATP the enzyme was in the E$_1$ state and therefore not susceptible to be phosphorylated by Pi. When ATP was added a Na$^+$-ATPase cycle started and the E$_2$ dephospho form, appearing after the dephosphorylation step, was susceptible to incorporate Pi. Therefore, to start with the problem the most logic assumption to us was to propose that the Na$^+$- and Na$^+$, K$^+$-ATPase cycles were practically identical. The model shown in Figure 2 is a simplified Albers-Post scheme (Albers, 1967) where the E$_1$P–E$_2$P transition was omitted. This was done because (i) there are no reliable values for the unidirectional rate constants, and (ii) it simplifies the resulting equations without altering the conclusions. As in the original Albers-Post, we included an E$_2$ form that can return to E$_1$ without or with binding ATP to a regulatory site. The unidirectional rate constants used for calculations, are:

f1	E$_1$+ATP → E$_1$ATP	0.025 ms^{-1} μM^{-1}
b1	E$_1$ATP → E$_1$+ATP	0.009 ms^{-1}
f2	E$_1$ATP → E$_2$ATP	1.95e^{-8} ms^{-1}
b2	E$_2$ATP → E$_1$ATP	0.054 ms^{-1}
f3	E$_2$ATP → E$_2$+ATP	0.16 ms^{-1}
b3	E$_2$+ATP → E$_2$ATP	0.0016 ms^{-1} μM^{-1}
f4	E$_2$ → E$_1$	0.1 ms^{-1}
b4	E$_1$ → E$_2$	0.00001 ms^{-1}
f5	E$_1$ATP → E$_2$P	0.199 ms^{-1}
f6(Na)	E$_2$P → E$_2$+Pi	0.0019 ms^{-1}
b6	E$_2$+Pi → E$_2$P	0.00006 ms^{-1} μM^{-1}

Their source and/or rationale can be found in Campos & Beaugé (1994). The true affinities (Ks) of E$_2$ for ATP and Pi were initially assumed to be 100 μM. However, the fitting of the data required a reduction of the Ks for Pi to 32 μM (by increasing b6). In Figure 2 there is a path leading to E$_2$P*. This addition, which does not alter the model, is needed to compute at the same time E$_2$P formed from ATP in the forward reaction and E$_2$P from inorganic phosphate (E$_2$P*). The concentration of P was always taken equal to zero, while that of P* represented the total concentration of inorganic phosphate in the media (cold plus radioactive). The scheme gives six simultaneous differential equation which were solved for pre- and/or steady state situations. That allowed us to obtain solutions for perturbations (for instances addition of inorganic phosphate) leading to new transient and steady state situations. Theoretical calculations of ATPase activity were done on the basis of the rate constant for E_2P breakdown (f6 = 114 min^{-1}) times the concentration of E_2P from ATP at steady state. Simulations and curve fittings were done with the Scop program (Simulation Resources, Berrien Springs, USA). As a check of the model and elected rate constants we obtained an excellent fitting of pre-steady state phosphorylation data from our own laboratory using ATP concentrations from 0.75 µM to 100 µM.

The experiments were performed with partially purified pig kidney enzyme as described by Campos & Beaugé (1994) either as a broken membranes (unsided) or incorporated into liposomes (sided) preparations. Temperature was 20°C. All incubation media contained 1.8 mM free Mg^{2+} and 0.1 mM EGTA; the pH was 7.4 with broken membranes and 7.0 with proteoliposomes. The total number of phosphorylating sites (E_{tot}) was estimated on the basis of the [^{32}P]Pi incorporation after 10 minutes incubation at 37°C in the presence of 1 mM [^{32}P]phosphate and 1 mM ouabain. The reader interested in other experimental procedures or theoretical details can find them in Campos & Beaugé (1994).

3.2. The Presence of Inorganic Phosphate during Na$^+$-ATPase Activity Uncovers the Low Affinity Regulatory ATP Site

The strategy was to simulate steady state turnover conditions and to measure actual ouabain-sensitive Na$^+$-ATPase activity as a function of [ATP] in the absence and presence of inorganic phosphate. Both types of data were then plotted in semilogarithmic scales (due to that large range in [ATP]) and fitted to a single michaelian (no Pi) and the sum of two michaelian (with Pi). The rationale for the whole approach, and the results obtained, go as follows. In the absence of external potassium the rate of dephosphorylation of E_2P is so slow (f6 = 0.019 ms^{-1}) that it becomes rate limiting. In addition, the rate constants for the E_1–E_2 transitions are similar without (f4 = 0.1 ms^{-1}) and with (b2 = 0.054 ms^{-1}) ATP bound to the enzyme; furthermore, their location in the cycle is away from the rate limiting step. Thus, it is of no surprise that even with two ATP binding sites with large affinity differences the ATP activation displays a single michaelian kinetics. This is what the model predicts (Fig. 3A) and what is actually observed.

(Fig. 3B). The predictions of the model are a K_{mATP} of 0.77 ± 0.0001 µM and a V_{max} of 0.18 ± 0.0002 µmoles Pi mg^{-1}. min^{-1}. The values obtained from the fit of the experimental data were a K_{mATP} of 0.22 ± 0.03µM and a V_{max} of 0.18 ± 0.005 µmoles Pi mg^{-1}. min^{-1}. Accordingly, to detect the low affinity ATP site the limiting step/s must be moved to the E_2 enzyme form. In the absence of potassium, one way to do that is to have a ligand binding to E_2 and taking that complex away from the hydrolytic path. The effect of that ligand would then be antagonized by ATP binding to the regulatory site. Inorganic phosphate appeared as a likely candidate acting as product inhibitor (Pedemonte & Beaugé, 1983) by forming the E_2P complex through the "back door phosphorylation" of the enzyme. The sum of two michaelians in predicted (Fig. 3A) and observed (Fig. 3B) when the ATP activation curve is determined in the presence of 3 mM inorganic phosphate. The predicted kinetic parameters were a K_{mATP1} of 0.029 ± 0.0001 µM, a K_{mATP2} of 449 ± 4 µM and a V_{max} of 0.18 ± 0.00002 µmoles Pi mg^{-1}. min^{-1}. From the fitting of the experimental points we obtained a K_{mATP1} of 0.09 ± 0.01 µM, a K_{mATP2} of 449 ± 4 µM and a V_{max} of 0.18 ± 0.00002 µmoles Pi mg^{-1}. min^{-1}. It is interesting that, although the fitting equations were arbitrarily chosen, upon the addition of 3 mM phosphate there is a reduction in K_{mATP1} which is fractionally identical in the experimental and theoretical curves.

Based on the above results, we should observe protection against Na$^+$-ATPase inhibition by inorganic phosphate by compounds that mimic ATP on the regulatory site. We tried two: AMP-PCP, an ATP analogue that does not phosphorylate but competes with

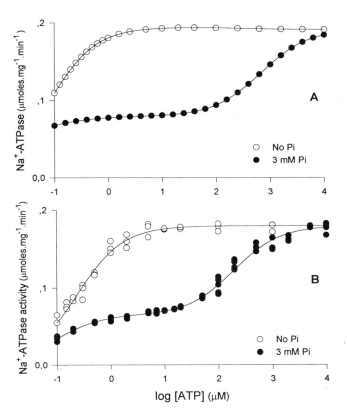

Figure 3. ATP stimulation of Na⁺-ATPase activity in the absence and presence of 3 mM inorganic phosphate. (A) Computed simulation on the basis of the model of Figure 2 and the rate constants given in the text. (B) Experimental data on the ouabain-sensitive ATP hydrolysis of Na⁺,K⁺-ATPase incubated at 20°C in media containing 150 mM NaCl, 30 mM Imidazole (pH 7.4 at 20°C), 1.8 mM free Mg²⁺ and variable concentrations of ATP, without (filled symbols) and with (open symbols) 3 mM inorganic phosphate. In (A) and (B) the lines through the points are the best fit to a single michaelian function (without Pi) or the sum of two michaelian (with 3 mM Pi). Taken from Campos and Beaugé, 1994 with permission.

ATP on the catalytic site (Rossi & Garrahan, 1989) and palmitoyl-Coenzyme A, (PCoA) a non ATP analogue which gets into the low affinity but does not bind to the catalytic domain. However, PCoA has a non specified inhibition of ATP hydrolysis when present at high concentrations (Huang et al., 1989). The results demonstrated that both compounds indeed protected against phosphate inhibition of Na⁺-ATPase activity (not shown).

3.3. Other Results Predicted by the Model

Apart from the basic expectation that no phosphorylation by Pi will be observed in the presence of Na⁺ and Mg²⁺ if there is no ATP (Post et al., 1975), there are other predictions from the model outlined in Figure 2. These are related to cases *when the enzyme is turning over in the Na⁺-ATPase modality*. They were experimentally verified and are summarized below:

a. Pre-steady and steady state levels of phosphorylation from inorganic phosphate:
   (i) For all phosphate concentrations, Pi incorporation proceeds with a $t_{1/2}$ of about 400 ms; i.e. a steady state incorporation is reached after 2 to 3 seconds; this is ten times slower than the values expected by simply applying the f6 and b6 rate constant *in the absence of Na⁺-ATPase turnover*.
b. As a function of the [Pi] the steady state levels of E-P formation follows a michaelian function with a K_{mPi} around 2000 µM; the observed value was 2020 ± 80 µM. Notice that this is 60–70 times larger than the 32 µM of the Ks for Pi.

An additional expectation is that at sufficiently high [Pi] all enzyme will be phosphorylated. The extrapolation of the experimental data gave an $E_2P^*_{max}$ of 0.99 ± 0.05 percent.

c. ATP should antagonize Pi incorporation. The model predicts, and the data substantiates, that all Pi incorporation is sensitive to ATP. The inhibition of phosphorylation as a function of [ATP], both theoretical and experimental, had an excellent fit to the equation

$$E_2P^* = E_2P^* (1 - ([ATP] / (K_m + [ATP])))$$

where E_2P^*, the fractional phosphoenzyme formed from Pi, can be fully inhibited by ATP. In this equation Km is the apparent affinity with which ATP prevents Pi incorporation. In simulations and experiments we consistently found this K_m two to three times larger that the Ks for ATP (see also Figure 4 in the next section).

3.4. The Effects of Na^+ on Phosphorylation by Pi during Na^+-ATPase Turnover

Working at 0°C Post et al. (1975) showed that, in the presence of Mg^{2+} and ATP, Na^+ antagonized Pi incorporation into the Na^+, K^+-ATPase. The mechanism for that effect was not investigated at that time. Also, it can not be deduced a priory from the kinetic model proposed by us. We started to reexamine the matter by studying, always at 20°C, the effects of two extreme Na^+ concentrations, 10 mM and 150 mM at constant and variable ionic strength and anion composition. The results showed that a fifteen fold increase in [Na^+] inhibited Pi incorporation to one half. This is a genuine Na^+ effect for it was not affected by ionic strength or anionic composition (Campos & Beaugé, 1994).

A dose response to Na^+ between 10 mM and 150 mM and at two concentrations of ATP, 2 μM and 100 μM, is illustrated in Figures 4A and B. In Fig. 4A the data points were fitted to the same equation utilized in ATP inhibition of Pi phosphorylation. The following results can be extracted from this Figure: (i) all Pi incorporation is inhibitable by Na^+. The [EP] vs [Na^+] relationship follows a *single site kinetics function* with a K_{mNa} of 65 ± 6 mM; (ii) potentiation between Na^+ and ATP is observed; and (iii) [ATP] has no influence of the apparent Na^+ affinity. We also followed the ATP inhibition of Pi incorporation at 10 mM and 150 mM Na^+. The data, fitted to the same equation, produces an exact reproduction of Figure 4A for (i) the ATP and Na^+ effects are additive, (ii) ATP inhibits along a single site function (see also Figure 4), and (iii) the K_{mATP} is independent of the Na^+ concentration.

The sidedness of the effects of Na^+ was investigated in liposomes with Na^+, K^+-ATPase incorporated under four conditions: (i) low (10 mM) Na^+_{cyt} – low (10 mM) Na^+_{ext}; (ii) high (200 mM) Na^+_{cyt} – low (10 mM) Na^+_{ext}; (iii) low (10 mM) Na^+_{cyt} – high (200 mM) Na^+_{ext}; (iv) high (200 mM) Na^+_{cyt} – high (200 mM) Na^+_{ext}. These results, which appear in Table 1 indicate that the Na^+ ions responsible for antagonizing Pi incorporation during Na^+-ATPase turnover act on extracellular sites. What about mechanism/s? One possibility is Na^+ binds to allosteric external sites with low affinity (Karlish et al., 1985; Beaugé, 1988) leading the enzyme to a conformation which does not accept Pi. On the other hand, it is known that, in a K^+-like action, Na^+ dephosphorylates the enzyme; interestingly this occurs with a $K_{0.5}$ about 33–40 mM (Glynn & Karlish, 1976; Beaugé & Glynn, 1979b) and it is likely to proceed through the E_2P - E_2PNa_2 - E_2Na_2 - E_1Na_2 - E_1 pathway. If E_2Na_2 is refractory to Pi, while E_2 is not, we could account for the results. It could be argued that an increase in the rate of E_2P breakdown leads to an increase in the E_2 form, susceptible to Pi. Actual calculations using our model show that

Figure 4. Combined effects of NaCl and ATP concentrations (at constant ionic strength) on the steady-state fractional Pi incorporation from inorganic phosphate during Na⁺-ATPase turnover at 20°C. After 1 sec preincubation in variable concentrations of NaCl and ATP, 30 mM Imidazole (pH 7.4 at 20°C) and 1.8 mM [Mg²⁺], 1 mM (final concentration) [³²P]Pi and enough MgCl₂ to keep [Mg²⁺] constant were added. The reaction was stopped after 3 sec of [³²P]Pi addition. In (A) and (B) the data points correspond to the experimental data obtained in duplicate and were simultaneously fitted to the equation given in the text. (A) The effects of different NaCl concentrations (up to 250 mM) at 10 mM and 100 mM ATP. Note: (i) at both ATP concentrations Na⁺ inhibition of Pi incorporation follows a single site kinetics with an apparent affinity of 64 ± 7 mM; (ii) all Pi incorporation is inhibitable by Na⁺; (iii) there is potentiation of Na⁺ and ATP inhibition; (iv) changing [ATP] from 10 μM to 100 μM does not affect the apparent affinity for Na⁺. (B) The effects of different ATP concentrations at 10 mM and 150 mM NaCl. Note: (i) at both NaCl concentrations ATP inhibition of Pi incorporation follows a single site kinetics with an apparent affinity of 128 ± 17 μM; (ii) all Pi incorporation is inhibitable by ATP; (iii) there is potentiation of Na⁺ and ATP actions; (iv) changing [NaCl] from 10 mM to 150 mM does not affect the apparent affinity for ATP. Taken from Campos and Beaugé, 1994 with permission.

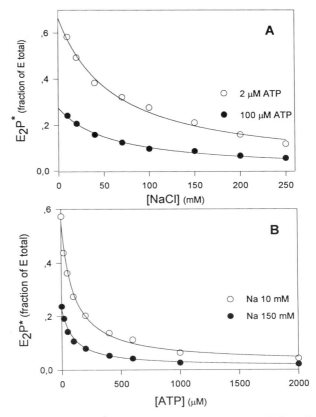

Table 1. Effects of intra and extracellular Na⁺ on the steady state levels of phosphorylation from Pi of Na⁺, K⁺-ATPase incorporated into liposomes during Na⁺-ATPase turnover at 20° with 2 μM ATP. After 2 sec preincubation in the absence and presence of 2 μM ATP, 1 mM [³²P]Pi at constant 1.8 mM [Mg²⁺] were added and the reaction was stopped 3 sec thereafter. The composition of solutions was as follows: (i) intracellular (extravesicular): 1.8 mM Mg²⁺; 30 mM imidazole (pH 7.4 at 20°C); 0.1 mM EGTA; 200 mM (low Na⁺) or none (high Na⁺) Tris.HCl; (ii) extracellular (intravesicular): the same as the intracellular except for the lack of Mg²⁺ ions. The total number of phosphorylating sites (E_{tot}) was estimated by [³²P]Pi incorporation in the same batches of liposomes 20 minutes after the addition of 2 mM digitoxigenin into the extravesicular solution

Cytoplasmic Na⁺ (mM)	Extracellular Na⁺ (mM)	ATP-dependent E-P (fraction of E_{tot})
10	10	0.459 ± 0.084
200	10	0.387 ± 0.042
10	200	0.119 ± 0.040
200	200	0.070 ± 0.027

The entries represent the mean ± s.e.m. of three different experiments. Taken from Campos and Beaugé, 1994 with permission.

even with a 10-fold increase in f6, a two-fold reduction in b6 leads to inhibition of Pi incorporation during Na^+-ATPase activity; i.e. E_2Na_2 does not accept Pi because a decrease in the "on" rate of phosphate binding with or without a concurrent increase in f6.

There are two additional questions regarding the Na^+ effects that had to be investigated further. On the one hand, it was suggested to us the possibility that what Na^+ ions actually do is to revert the ATPase cycle leading to ATP synthesis, not hydrolysis. The idea came about because in other ATPases even micromolar concentrations of ADP (which could be present as contaminant in our system) can trigger the reverse reaction (de Meis, personal communication). To explore this possibility we decided to run in parallel experiments of Pi incorporation and ATP hydrolysis as a function of [Na^+]. On the other hand, it is known that, at concentrations below 10 mM (maximal effect at 2.5–5.0 mM) external Na^+ can block E_2P dephosphorylation thus inhibiting Na^+-ATPase, uncoupled Na^+ efflux and ATP-ADP exchange reactions (Glynn & Karlish, 1976; Beaugé & Glynn, 1979b; Garrahan & Glynn, 1967; Beaugé & Campos, 1986). If our scheme for Na^+-ATPase is correct, another consequence of slowing E_2P dephosphorylation must be a reduction of steady state [E_2] and Pi incorporation. This was also explored in the same experiments. The results of one of them is illustrated in figure 5 and can be summarized as follows: (i) As [Na^+] is increased above 10 mM, the reduction of Pi incorporation is accompanied by and increase in ATP hydrolysis. This shows that if there any reversal of the cycle that is negligible. (ii) At [Na^+] below 10 mM the classical intermediate affinity inhibition of Na^+-ATPase is observed (Glynn & Karlish, 1976; Beaugé & Glynn, 1979b; Garrahan & Glynn, 1967; Beaugé & Campos, 1986). This coincides with a marked inhibition,

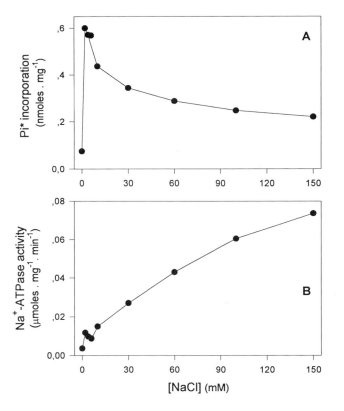

Figure 5. Simultaneous estimation of Pi incorporation and Na^+-ATPase activity of Na^+, K^+-ATPase as a function of the Na^+ concentration. The incubation solutions contained 40 mM Imidazol (pH 7.4 at 20°C), 150 - [NaCl] Tris-HCl (pH 7.4 at 20°C), 1.8 mM free Mg^{2+}, 0.1 mM EGTA and variable [NaCl]. In Pi incorporation experiments there was 1 mM [^{32}P]Pi and 10 μM cold ATP. In ATPase experiments there was 1 mM cold Pi and 10 μM [^{32}P]P-γ-ATP. We define as Na^+-ATPase activity the difference in ATP hydrolysis determined in the absence and presence of 10^{-4} M ouabain. Temperature was 20°C. For more details see legend to Figure 3. (Unpublished experiment of Pezza, Campos, and Beaugé).

about one half, of Pi incorporation. Therefore, the antagonism of external Na⁺ to the incorporation of Pi during Na⁺-ATPase turnover has two mechanisms: below 10 mM it does so by slowing the E2P breakdown; above 10 mM, and as it reverts the inhibition of E_2P hydrolysis, it leads to the formation of E_2Na_2 which is refractory to Pi phosphorylation.

4. THE INTERPLAY OF Na⁺ AND ATP ON VANADATE INHIBITION OF Na⁺-ATPase

Vanadate is a potent inhibitor of the Na⁺,K⁺-ATPase activity and the coupled Na⁺-K⁺ transport (Beaugé, 1988; Beaugé et al., 1980). Actually, in red cells and squid axons treated with vanadate a reversal K⁺-free effect is observed; i.e. there is an increase, instead of a decrease, in Na⁺ efflux following removal of extracellular K⁺ (Beaugé, 1988; Beaugé, 1979). A reversal in the K⁺-free effect is also found when the intracellular ATP concentration is very low. On the other hand, in K⁺-free Na⁺-containing solutions vanadate does not bind to the non turning partially purified Na⁺,K⁺-ATPase (Beaugé, 1988; Karlish et al.; 1979). If vanadate indeed binds to the phosphate site (Cantley et al. 1978), it should be expected that its ability to inhibit the Na⁺-ATPase activity would depend on the ATP and Na⁺ concentration; from the results shown above one would anticipate that high [ATP] and [Na⁺] will be the most unfavorable, while low [ATP] and [Na⁺] the most favorable for vanadate inhibition. One of these experiments is illustrated in figure 6 while Table 2 summarizes the $K_{0.5}$'s values for vanadate in each condition. The analysis of the data shows that the most adverse condition for vanadate inhibition is in fact the combination of high [ATP] and [Na⁺]. In this regard, Na⁺ appears to be more efficient than ATP. A one thousand fold increase in ATP concentration (from 3 µM to 3000 µM) increases the $K_{0.5V}$ by

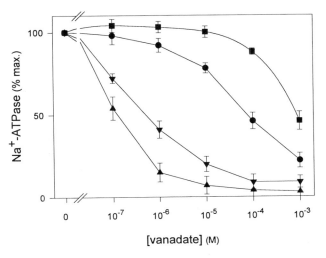

[vanadate] (M)

Figure 6. Dose response curves for vanadate inhibition of Na⁺-ATPase activity of partially pig kidney Na⁺, K⁺-ATPase as a function of Na⁺ and ATP concentrations. The incubation solutions contained 40 mM Imidazol (pH 7.4 at 20°C), 150 - [NaCl] Tris-HCl (pH 7.4 at 20°C), 1.8 mM free Mg²⁺, 0.1 mM EGTA, 10 mM or 150 mM NaCl, 3 µM or 3000 µM ATP and variable vanadate concentrations. We define as Na⁺-ATPase activity the difference in ATP hydrolysis determined in the absence and presence of 10^{-4} M ouabain. The different conditions explored were: 3000 µM ATP–150 mM NaCl (■); 3 µM ATP–150 mM NaCl (●); 3000 µM ATP–10 mM NaCl (▲); and 3 µM ATP–10 mM NaCl (▼). Temperature was 20°C. For more details see legend to Fig. 3. (Unpublished experiment of Pezza, Campos and Beaugé).

Table 2. Effects of Na^+ and ATP concentrations on
the apparent affinities for vanadate inhibition
of the Na^+-ATPase activity of the Na^+, K^+-ATPase

[ATP] (μM)	[Na$^+$] (mM)	K_{i50V} (M)
3	10	1.0×10^{-7}
3	150	9.0×10^{-5}
3000	10	4.7×10^{-7}
3000	150	7.0×10^{-4}

Data taken from Figure 6.

4.7-fold at 10 mM Na^+ and 7.8-fold at 150 mM Na^+. Conversely, a 15-fold rise in Na^+ concentration (from 10 mM to 150 mM) leads to a 900 times and 1500 times increase in the $K_{0.5V}$ at 3 μM and 3000 μM ATP respectively. Indeed, the antagonizing effects of Na^+ on vanadate inhibition are likely to involve mechanisms other than just the formation of the refractory E_2Na_2 state. Actually, the data can be interpreted by the proposed extracellular sites which, when occupied by Na^+, reduce the intracellular vanadate binding affinity (Beaugé, 1988). This interpretation makes sense in view of the difference between phosphate and vanadate effects on the Na^+,K^+-ATPase activity. In the presence of external Na^+, phosphate inhibition is not influenced by extracellular K^+, whereas that due to vanadate increases together with external K^+ concentration. In the absence of extracellular Na^+ the fractional phosphate and vanadate inhibitions are both insensitive to external K^+ (Beaugé, 1988; Pezza, Campos, and Beaugé, unpublished). In other words, the data is consistent with the binding of vanadate to a site in E_2 while E_2Na_2 is refractory to that binding; but it seems that more is needed to fully account for its interactions with Na^+ and ATP.

5. CONCLUSIONS

The experiments described in this work provide convincing evidence that, even in the absence of external K^+, the cycle of ATP hydrolysis by the Na^+, K^+-ATPase enzyme go through an E_2 conformation that has the same low affinity ATP site observed in the $E_2(K_2)$ "occluding" state. In turn this means that, if not all, at least most of the steps following enzyme dephosphorylation are the same with and without external K^+. In addition, the data stress the similarities and differences between phosphate and vanadate inhibition of the Na^+, K^+-ATPase activity.

6. SUMMARY

We studied the reactivity of the E_2 form of the Na^+, K^+-ATPase (EC 3.6.1.37), the Na^+ gradient forming enzyme, by following the sensitivity of the Na^+-ATPase activity to inorganic phosphate (Pi) and vanadate (V) and the phosphorylation of the enzyme from Pi during Na^+-ATPase turnover at 20°C. For theoretical calculations we used the Albers-Post scheme assuming that, even with no K^+, E_2 can return to E_1 without or with ATP bound to the regulatory site. The experiments were carried out with partially purified pig kidney Na^+, K^+-ATPase. The model, using rate constants obtained from the literature, predicts: (i) the Na^+-ATPase activity shows michaelian kinetics with high ATP affinity in the absence of Pi and the sum of two michaelians, with high and low affinity ATP sites, in its presence; (ii) Pi incorporation is reduced by ATP acting with low affinity; (iii) the apparent affinity for Pi is substantially lower that its Kd. The results agreed with the predictions. In addition, Na$^+$ and ATP fully inhibited Pi incorporation; although there was potentiation the Ki's were independent of the other ligand concentration. At [Na$^+$] above 10 mM the reduction of Pi incorporation was accompanied by an increase in ATP hydrolysis; this rules out any reversal of the cycle. At [Na$^+$] below 10 mM the intermediate affinity inhibition of Na$^+$-ATPase and E$_2$P breakdown coincided with inhibition of one half of Pi incorporation. Therefore, during Na$^+$-ATPase turnover external Na$^+$ antagonizes Pi incorporation via two mechanisms: by slowing the E$_2$P breakdown below 10 mM and by leading to E$_2$Na$_2$ intermediate refractory to Pi above 10 mM. Regarding V, the data showed that the effects of ATP and Na$^+$ are quantitatively different from those with Pi. Therefore, the Na$^+$-V antagonism requires mechanisms other than the formation of the refractory E$_2$Na$_2$ state. These results indicate that, in the absence of external K$^+$, the cycle of ATP hydrolysis by the Na$^+$, K$^+$-ATPase goes through an E$_2$ conformation that has the same low affinity ATP site observed in the E$_2$(K$_2$) "occluding" state.

7. ACKNOWLEDGMENTS

This work was supported by Grants from CONICET, CONICOR, Volkswagen-Stiftung I / 72 122 and Fundación Andes (C-1277/9). M.C. and L.B. are established investigators from CONICET and R.P. is a research student.

8. REFERENCES

Albers, R.W. (1967) Biochemical aspects of active transport. Ann. Rev. Biochem. 36:727–756.

Aronson, P.S. (1985) Kinetic properties of the plasma membrane Na$^+$-H$^+$ exchanger. Ann. Rev. Physiol. 47:545–560

Beaugé, L.A. (1979) Vanadate-potassium interactions in the inhibition of Na$^+$,K$^+$-ATPase. In Na$^+$,K$^+$-ATPase: Structure and Kinetics, J.C. Skou and J.G. Norby, eds. pp. 373–387, Academic Press, London.

Beaugé, L.A. (1988) Inhibition of translocation reactions by vanadate. Methods Enzymol. 156:251–267.

Beaugé, L.A. and Campos, M.A. (1986) Effects of mono and divalent cations on total and partial reactions catalyses by pig kidney Na$^+$,K$^+$-ATPase. J. Physiol. 375:1–25.

Beaugé, L., Berberián, G. and Campos, M. (1990) Potassium occlusion in relation to cation translocation and catalytic properties of Na$^+$,K$^+$-ATPase. In Regulation of Potassium Transport across Biological Membranes, Reuss, L., Russell, J.M., and Szabo, G., eds. pp 29–63, The University of Texas Press, Austin, Texas.

Beaugé, L.A., Cavieres, J.J., Glynn, I.M. and Grantham, J.J. (1980) The effects of vanadate on the fluxes of sodium and potassium ions through the sodium pump. J. Physiol. 301:7–23.

Beaugé, L.A., and Glynn, I.M. (1979a) Occlusion of K$^+$ ions in the unphosphorylated sodium pump. Nature 280:510–512.

Beaugé, L.A. and Glynn, I.M. (1979b) Sodium ions, acting at high-affinity extracellular sites, inhibit sodium ATPase activity of the sodium pump by slowing dephosphorylation. J. Physiol. 289:17–31.

Campos, M. and Beaugé, L. (1994) Na$^+$-ATPase activity of the Na$^+$,K$^+$-ATPase. Reactivity of the E2 form during Na$^+$-ATPase turnover, J. Biol. Chem. 269:18028–18036.

Cantley, L.C., Cantley, L.G. and Josephson, L. (1978) A characterization of vanadate interactions with the Na$^+$,K$^+$-ATPase. J. Biol. Chem. 253:7361–7368.

Cornelius, F. (1991) The kinetics of uncoupled fluxes in reconstitutes vesicles. in The Sodium Pump: Structure, Mechanism and Regulation, J.H. Kaplan and P. De Weer, eds., pp 267–280, The Rockefeller University Press, N.Y.

Forbush, B. (1988) Overview: Occluded ions and Na$^+$,K$^+$-ATPase, in The Sodium Pump, in The Sodium Pump, Part A: Molecular aspects. Skou, J.C., Norby, J.G., Maunsbach, A.B. and Esmann, M., eds. pp 229–248, Alan R. Liss, Inc., New York.

Froehlich, J.P. and Fendler, K. (1991) The partial reactions of the Na$^+$- and Na$^+$+K$^+$-activated adenosine triphosphatases, in The Sodium Pump: Structure, Mechanism, and Regulation, Kaplan, J.H., and DeWeer, P., eds. pp 227–247, The Rockefeller University Press, New York.

Garrahan, P.J., and Glynn, I.M. (1967) Factors affecting the relative magnitudes of the sodium-potassium and sodium-sodium exchanges catalyzed by the sodium pump. J. Physiol. 192:189–216.

Glynn, I.M. (1985) The Na^+,K^+-transporting adenosine triphosphatase. in The enzymes of biological membranes, 2nd. edn, vol. 3, Martonosi, A.N., ed. pp 35–114, Plenum Press, New York.

Glynn, I.M., and Karlish, S.J.D. (1976) ATP hydrolysis associated with uncoupled sodium efflux through the sodium pump: evidence for allosteric effects of intracellular ATP and extracellular sodium. J. Physiol. 256:465–496.

Glynn, I.M., and Karlish, S.J.D. (1990) Occluded cations in active transport. Annu. Rev. Biochem. 59:171–205.

Huang, W., Wang, Y. and Askari, A. (1989) Mechanism of the control of Na^+,K^+-ATPase by long-chain Acyl Coenzyme A. J. Biol. Chem. 264:2605–2608.

Karlish, S. J. D., Beaugé, L.A., and Glynn, I. M. (1979) Vanadate inhibits Na^+,K^+-ATPase by blocking a conformational change of the unphosphorylated form. Nature 282:333–335

Karlish, S.J.D., Rephaeli, A., and Stein, W.D. (1985) Transmembrane modulation of cation transport by the Na^+,K^+-pump. in The Sodium Pump, Glynn, I., and Ellory, C., eds. pp 485–499, The Company of Biologists Ltd., Cambridge.

Norby, J.G., and Klodos I. (1988) The phosphointermediates of Na^+,K^+-ATPase, in The Sodium Pump, Part A: Molecular aspects, Skou, J.C., Norby, J.G., Maunsbach, A.B., and Esmann, M., eds. pp 249–270. Alan R. Liss, Inc., New York.

Pedemonte, C.H., and Beaugé L.A. (1983) Inhibition of (Na^+,K^+)-ATPase by magnesium ions and inorganic phosphate and release of these ligands in the cycles of ATP hydrolysis, Biochim. Biophys. Acta 748:245–253.

Post, R.L., Toda, G., and Rogers, F.N. (1975) Phosphorylation by inorganic phosphate of sodium plus potassium ion transport adenosine triphosphatase. J. Biol. Chem. 250:691–701.

Rossi, R., and Garrahan, P.J. (1989) Steady-state kinetic analysis of the Na^+,K^+-TPase. The effects of adenosine 5'[beta,gamma-methylene]triphosphate on substrate kinetics, Biochim. Biophys. Acta 981:85–94.

Tanford, C. (1983) Mechanism of free energy coupling in active transport, Ann. Rev. Biochem. 52:379–409

Ullrich, K.J. (1979) Sugar, amino acid, and Na^+ cotransport in the proximal tubule, Ann. Rev. Physiol., 41:181–195

10

MODULATION OF P-GLYCOPROTEIN ON TUMOUR CELLS

Monique Orind,[1] Karen Wagner-Souza,[1] Raquel C. Maia,[2] and Vivian M. Rumjanek[1*]

[1]Laboratório de Imunologia Tumoral
Instituto de Biofísica Carlos Chagas Filho
Universidade Federal do Rio de Janeiro
Rio de Janeiro, Brazil
[2]Serviço de Hematologia e Centro de Pesquisa Básica
Instituto Nacional de Cancer
Rio de Janeiro, Brazil

1. INTRODUCTION

A serious problem in cancer chemotherapy involves the generation of multidrug resistance (MDR). This phenomenon represents cross-resistance among a number of drugs, unrelated structurally or functionally, and is one of the major reasons for chemotherapy failure. MDR is associated to the overexpression of the *mdr1* gene which encodes a 170kDa plasma membrane glycoprotein known as P glycoprotein (Pgp) (Chen et al., 1986; Juliano and Ling, 1976; Riordan et al., 1985) and it has been demonstrated that overexpression of this protein is sufficient to confer cellular resistance (Ueda et al. 1987). Pgp functions as an ATP-dependent efflux pump capable of extruding antineoplastic agents to the outside of the cell reducing their intracellular levels. Structurally, Pgp is a transmembrane protein arranged in two homologous halves, each half containing an ATP binding site facing the cytoplasm, and with twelve hydrophobic regions forming the transmembrane loops (Gottesman and Pastan, 1993).

Pgp activity can be regulated by protein kinase C phosphorylation and it has been proposed that a number of substances known as MDR-reversing or modulatory agents may exert their effect by inducing Pgp phosphorylation and producing conformational changes capable of affecting the functional activity of this pump (Fine et al., 1985; Hamada et al., 1987).

* Corresponding author.

Calcium and Cellular Metabolism: Transport and Regulation, edited by Sotelo and Benech.
Plenum Press, New York, 1997

2. EXPRESSION OF Pgp IN HAEMATOLOGICAL MALIGNANCIES

High levels of Pgp expression can be detected in normal tissues like kidney, liver, colon, adrenal gland and the endothelial cells in brain and testes (Thiebaut et al., 1987); whereas in haematopoietic cells the expression is low (Chaudhay and Ronison, 1991; Drach et al., 1992). This does not preclude, however, the fact that the MDR phenotype has been found in a number of haemathological malignancies where it correlates with an unfavorable prognosis (Nooter and Sonneveld, 1994). Both untreated and relapsed leukaemias may express P-glycoprotein, although intrinsic expression, i.e. that seen in untreated patients is less common in haematological malignancies when compared to that reported for other types of human cancer, for example renal and coloretal (Pastan and Gottesman, 1991). In our experience more than 50% of patients with haematological malignancies at the Instituto Nacional de Cancer in Rio de Janeiro possessed the MDR phenotype.

3. DETECTION OF THE MDR PHENOTYPE

Many methods have been used to analyze and detect multidrug resistance and the advantages and disadvantages of each method have been thoroughly analyzed by Chan et al. (1994). Some are based on the detection of gene expression, others on the presence of P-glycoprotein on the cell membrane with the use of antibodies and others on the functional assay of fluorescent dye extrusion that mimics drug exclusion (Neyfake, 1988). Another indication that the mechanism of resistance involves Pgp function is through the use of reversors or modulatory agents, substances capable of inhibiting the activity of the pump (Ford and Hait, 1990). A problem with the latter system is that samples from individual patients may respond differently to the various modulatory agents tested, some samples failing completely to respond to a given agent while being modulated by another (Fig.1). This suggests that small structural differences exist among Pgp molecules and they might direct the response to chemotherapeutic drugs and the various modulatory agents (Gottesman and Pastan, 1993).

4. CHARACTERIZATION OF A VINCRISTINE-RESISTANT CELL LINE DERIVED FROM K562 ERYTHROLEUKAEMIA

To avoid the problem of the variabilty seen when working with patient samples, this laboratory selected a vincristine-resistant subline, derived from the erythroleukaemia parental line K562, by gradually exposing these cells to increasing amounts of the drug using the technique described by Tsuruo and co-workers (1983). These cells are capable of growing equally well in normal medium or medium containing 60nM of vincristine, a concentration 20 times higher than the one necessary to kill the original parental line (Rumjanek et al. 1994). These vincristine-resistant cells, named by us K562-Lucena 1, share many of the characteristics of MDR cells: they are resistant to a number of anthracyclines (Maia et al. 1996), express Pgp on their surfaces as evidenced by immunocytochemistry using the JSB-1 monoclonal antibody and are capable of extruding the fluorescent dye rhodamine 123 (Rho123). Furthermore, modulation of the pump activity by the classical MDR modulatory agents such as verapamil (VRP), trifluoperazine (TFP) and and cyclosporin A (CS-A) , could also be observed in these cells (Fig. 2).

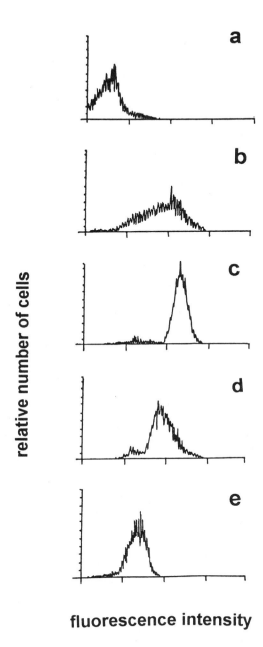

Figure 1. Response on leukaemic cells to modulatory agents. The amount of rhodamine within the cells was determined by flow cytometry as described in Maia et al. (1996). Cells were pre-loaded with rhodamine 123 (Rho123) for 45min at 37°C and left to extrude the dye in dye free medium for 45min at 37°C. Fluorescence profile of: a- Cells not exposed to Rho123; b-Cells exposed to Rho123; c-Cells exposed to Rho123 in the presence of 200μg/ml CS-A; d- Cells exposed to Rho123 in the presence of 5μM VRP; e- Cells exposed to Rho123 in the presence of 5μM TFP.

5. HEPARIN AS A MDR-MODULATORY AGENT

Not only VRP and TFP, but several other compounds known to interfere with rodhamine extrusion are capable of affecting intracellular calcium levels, a number of them, such as quinine, propranolol, phenotiazines, local anesthesics, by directly modifying the sarcoplasmic/endoplasmic reticulum Ca^{2+}ATPases (de Meis, 1991; de Meis et al., 1996; Gattas

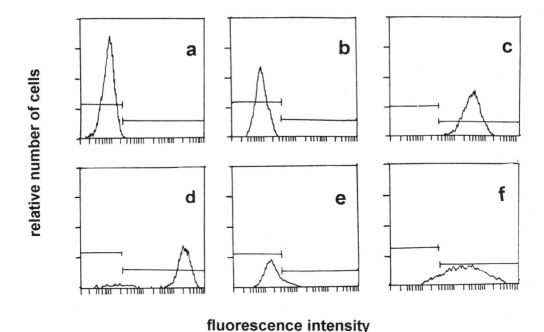

fluorescence intensity

Figure 2. Effect of verapamil (VRP) and cyclosporin A (CS-A) on rhodamine123 (Rho123) extrusion by K562-Lucena 1 cells (MDR-cell line). See legend of Fig.1 for details. Fluorescence profile of: a- K562 cells not exposed to Rho123; d- K562 cells pre-loaded with Rho123; b- K562-Lucena 1 cells not exposed to Rho123; c- K562-Lucena 1 cells pre-loaded with Rho123; d- K562-Lucena 1 cells pre-loaded with Rho123 in the presence of 5μM VRP; f- K562-Lucena 1 cells pre-loaded with Rho123 in the presence of 400ng/ml CS-A.

and de Meis, 1978). Using an inverse approach, our laboratory investigated the possibility that heparin, known to uncouple the Ca^{2+} pump (de Meis and Suzano, 1994) and to compete with IP3 for its receptors, might function as a MDR reversor. Despite producing a smaller response when compared to VRP or CS-A, heparin could modulate the functional activity of Pgp in K562-Lucena 1 and partially inhibited dye extrusion (Fig. 3b and 3c). Similar results were observed when samples from leukaemic patients were used (Maia et al., 1996).

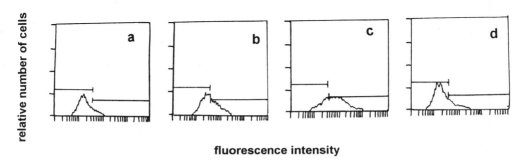

fluorescence intensity

Figure 3. Effect of heparin and thapsigargin (TG) on rhodamine 123 (Rho 123) extrusion by K562-Lucena 1 cells (MDR-cell line). See legend of Fig.1 for details. Fluorescence profile of: a- K562-Lucena 1 cells pre-loaded with Rho123; b- K562-Lucena 1 cells pre-loaded with Rho123 in the presence of 10μg/ml heparin; c- K562-Lucena 1 cells pre-loaded with Rho123 in the presence of 100μg/ml heparin; d- K562-Lucena 1 cells pre-loaded with Rho123 in the presence of 50nM TG.

Table 1. Effect of thapsigargin (TG) on tumour cell growth[a]

Cell type	VCR (nM)	TG (nM)	VRP (μM)	Cell number (×10⁴/ml)
K562	0	0	0	50.0
K562	60	0	0	1.5
Lucena 1	0	0	0	21.3
Lucena 1	60	0	0	22.3
Lucena 1	0	0	5	30.0
Lucena 1	60	50	0	21.0
Lucena 1	60	0	5	1.5

[a]Tumour cells, at an initial concentration of 2×10^4/ml were cultured for 7 days in the presence or absence of the various drugs, vincristine (VCR), thapsigargin (TG) and verapamil (VRP). At the end of the culture cell number and viability were assessed by phase microscopy.

6. EFFECT OF THAPSIGARGIN ON RESISTANT CELL LINES

It has been suggested that MDR modulatory agents inhibit drug efflux by competing with their binding site. In the case of heparin, the mode of action has not been investigated and could result from competition; from directly uncoupling the Pgp; or from a change in cytosolic calcium levels. The impact that changes in the cytosolic calcium levels could exert on Pgp mediated resistance was investigated using thapsigargin (TG). TG is a sesquiterpene lactone known to specifically inhibit sarcoplasmic/endoplasmic reticulum Ca^{2+}ATPases (Sagara et al., 1992; Thastrup et al., 1990), and to produce a rise in Ca^{2+} concentration in the cytoplasm as a result of emptying the Ca^{2+} storage pools followed by the entry of extracellular Ca^{2+} (Putney 1990). TG by itself or combined with vincristine did not affect cell growth of the MDR cell line (Table 1) despite its well known anti-proliferative action (Waldron et al., 1994).

Similarly, TG did not modify Rho123 dye exclusion in our resistant cells (Fig. 3d).

When the effect of TG was further analyzed using Fura-2 to measure intracellular Ca^{2+} levels, no Ca^{2+} accumulation could be detected in the cytoplasm of 562-Lucena 1 cell

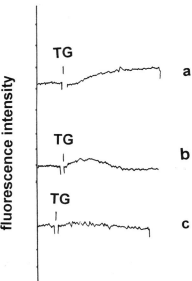

Figure 4. Effect of thapsigargin (TG) on Ca^{2+} mobilization. Levels of cytosolic free calcium were determined in a fluorescence spectrophotometer in cells previously loaded with Fura-2 in the presence of probenecid to inhibit organelle sequestration, based on the method described by di Virgilio et al. (1990). All preparations were exposed to 50nM TG. a- Peripheral blood mononuclear cells; b- K562 cells; c- K562-Lucena 1 cells (MDR-cell line).

lines (Fig. 4). This result is in agreement with what has been described by Hussain and co-workers (Gutheil et al., 1994, Hussain et al., 1995) who, using a different system and a number of different MDR cell lines, found a correlation between Pgp content and TG resistance. They, however, were capable of reversing this effect using inhibitors of Pgp function and suggested that TG might be a substrate for Pgp although this could not account for all the effect observed. Cells made resistant to TG show a MDR phenotype (Gutheil et al., 1994) and an overexpression of various Ca^{2+} transport ATPases (Hussain et al., 1995). It is possible that the pressure involved in the development of multidrug resistance might lead to the de-repression of a number of other ATPases in parallel and this might offer the tumour cells some survival advantages.

7. MDR-MODULATORY AGENTS AND NATURAL KILLER CELL FUNCTION

In an attempt to investigate this point, we looked at the effect of TG on natural killer (NK) cells that naturally express Pgp on their membrane (Kobayashi et al., 1994, Willisch et al., 1993). NK cells are cytolytic to a variety of tumour cells without the need of previous immunization and seem to play a role in tumour destruction and prevention of metastasis (Herberman and Ortaldo, 1981). The activity of these cells was extremely sensitive to TG (Table 2) being unlikely that the drug was being extruded, despite the fact that these cells are quite capable of extruding rhodamine 123.

The different responses depicted by tumour cells expressing Pgp and normal cells expressing the same glycoprotein, suggest that tumour cells selected for their MDR phenotype respond in a particular way because their internal enviroment is different.

8. SUMMARY

The MDR phenomenon has been associated with an energy-dependent mechanism mediated by a tranporter, belonging to the ATP-binding cassette (ABC) family of traffic ATPases, and known as P-glycoprotein (Pgp). The functional activity of Pgp can be assessed *in vitro* using a fluorescent dye extrusion test and may be modulated using MDR-reversors or modulatory agents. Most of these agents are capable of affecting cytosolic

Table 2. Effect of thapsigargin (TG) on natural killer (NK) cytotoxic activity[a]

Treatment	% Cytotoxicity		
	100:1*	50:1*	25:1*
None	38.5	26.9	6.7
5 nM TG	23.1	15.2	2.8
50 nM TG	16.6	8.4	2.0

*effector cell : target cell ratio

[a]Different concentrations of TG were added at the time of the assay. Cytotoxicity against K562 cells was determined by the ^{51}Cr release assay as in de LaRocque et al. (1995). Despite the fact that the physiological significance of Pgp expression by NK cells is still unknown, there are some evidences that it might be related to the cytolytic activity of these cells. A number of MDR-reversing agents such as VRP, metofoline, quinidine and reserpine (Chong et al., 1993; Klimeki et al., 1995) as well as TFP (de LaRocque et al., 1995) are capable of blocking NK cytotoxic activity. CSA and a non-immunossupressive derivative PSC although very efficient in reversing Rho123 efflux were less potent in inhibiting NK cytotoxicity (diPadova, 1989; Klimeki et al., 1995). It is quite clear that the relation between Pgp function and NK cytolysis needs to be better understood as a combination of the use of MDR-modulatory agents and immunotherapy is, potentially, a treatment modality for MDR tumours.

calcium levels but the relevance of this fact for MDR modulation is not clear. In the present work, using samples from leukaemic patients, it was possible to observe some variability among the responses of a given individual to the modulatory agents tested (verapamil, cyclosporin A and trifluoperazine). To be able to better analyze the regulation of the efflux pump, we established a vincristine-resistant cell line with all the characteristics of a MDR cell. Using this cell line it was observed that heparin, known to uncouple the Ca^{2+}ATPase, was capable of partially inhibiting rodamine 123 efflux. On the other hand, thapsigargin, an inhibitor of endoplasmic reticulum Ca^{2+}ATPase had no effect on the resistant cell line nor induced calcium mobilization in these cells. The possibility that thapsigargin is a substrate for Pgp and is therefore being extruded seems unlikely as this substance inhibited natural killer cytotoxicity, and this cell also expresses Pgp on the surface. The fact that the cell responsible for the antimetastatic function also expresses Pgp should be taken into consideration when using modulatory agents *in vivo*.

9. ACKNOWLEDGMENTS

This work was supported by Conselho Nacional de Desenvolvimento Científico e Tecnológico (CNPq-Brazil), FINEP and Fundação Ary Frauzino.

10. REFERENCES

Chan, H.S.L., DeBoer, G., Thorner, P.S., Haddad, G., Gallie, B.L. & Ling, V. (1994). Multidrug resistance. Clinical opportunities in diagnosis and circumvention. Hematol. Oncol. Clinics North America 8:383–410.

Chaudhay, P.M. & Ronison, I.B. (1991). Expression and activity of P-glycoprotein, a multidrug efflux pump, in human hematopoietic stem cells. Cell 66:85–94.

Chen, C.J., Chin, J.E., Ueda, K., Clark, D.P., Pastan, I., Gottesman, M.M. & Roninson, I.B. (1986). Internal duplication and homology with bacterial transport proteins in the *mdr-1* (P-glycoprotein) gene from multidrug-resistant human cells. Cell 47:381–389.

Chong, A.S.F., Markham, H.M., Gebel, S.D., Bines, S.D., & Coon, J.S. (1993). Diverse multidrug-resistance-modification agents inhibit cytolytic activity of natural killer cells. Cancer Immunol. Immunother. 36:133–139.

de LaRocque, L., Campos, M.M., Olej, B., Castilho, F., Mediano, I.F. & Rumjanek, V.M. (1995). Inhibition of human LAK-cell activity by the anti-depressant trifluoperazine. Immunopharmacol. 29:1–10, 1995.

de Meis, L. (1991). Fast efflux of Ca2+ mediated by the sarcoplasmic reticulum Ca2+ ATPase. J. Biol. Chem. 266:5736–5742.

de Meis, L. & Suzano, V.A. (1994). Uncoupling of muscle and blood platalets Ca^{2+} Transport ATPase by heparin: regulation by K^+. J. Biol. Chem. 269:14525- 14529.

de Meis, L., Wolosker, H. & Ergelender, S. (1996). Regulation of the Channel function of Ca2+ ATPase. Biochim. Biophys. Acta , *in press*.

di Padova, F.E. (1989). Pharmacology of cyclosporine: V- Pharmacological effects on immune function: in vitro studies. Pharmacol. Rev. 41:374–405.

di Virgilio, F., Steinberg, T.H. & Silverstein, S.C. (1990). Inhibition of Fura-2 sequestration and secretion with organic anion transport blockers. Cell Calcium 11:57–62.

Drach, D., Zhao, S., Drach, J., Mahadevia, R., Gattringer, C., Huber, H. & Andreeff, M. (1992). Subpopulations of normal peripheral blood and bone marrow cells express a functional multidrug resistant phenotype. Blood 80:2729–2735.

Fine, R.L., Patel, J., Allegra, C.J., Curt, G.A., Cowan, K.H., Ozols, R.F., Lippman, M.E., McDevitt, R. & Chabner, B.A. (1985). Increased phosphorylation of a 20,000 MW protein in pleiotropic drug-resistant MCF human breast cancer lines. Proc. Am. Assoc. Cancer. Res. 26:345.

Ford, J.M. & Hait, W.N. (1990). Pharmacology of drugs that alter multidrug resistance in cancer. Pharmacol Rev 42:155–199.

Gattas, C.R. & de Meis, L. (1978). The mechanism by which quinine inhibits the Ca2+ transport of sarcoplasmic reticulum. Biochem. Pharmacol. 27:539–545.

Gottesman, M.M. & Pastan, Y. (1993). Biochemistry of multidrug resistance mediated by the multidrug transporter. Annu. Rev. Biochem. 62:385–427.

Gutheil, J.C., Hart, S.R., Belani, C.P., Melera, P.W. & Hussain, A. (1994). Alterations in Ca^{2+} transport ATPase and P-glycoprotein expression can mediate resistance to thapsigargin. J. Biol. Chem. 269:7976–7981.

Hamada, H., Hagiwara, K.I., Nakajima, T. & Tsuruo, T. (1987). Phosphorylation of the Mr 170,000 to 180,000 glycoprotein specific to multidrug-resistant tumor cells: effect of verapamil, trifluoperazine, and phorbol esters. Cancer. Res. 47:2860–2865.

Herberman, R.B. & Ortaldo, J.R. (1981). Natural killer cells: their role in defenses against disease. Science 214, 24–30.

Hussain, A., Garnett, C., Klein, M.G., Tsai-Wu, J.J., Schneider, M.F. & Inesi, G. (1995). Direct involvement of intracellular Ca^{2+} Transport ATPase in the development of thapsigargin resistance by chinese hamster lung fibroblasts. J. Biol. Chem. 270:12140–12146.

Juliano, R.L. & Ling, V. (1976). A surface glycoprotein modulating drug permeability in chinese hamster ovary cell mutants. Biochim. Biophys. Acta 455:152–162.

Klimeki, W.T., Taylor, C.W. & Dalton, W.S. (1995). Inhibition of cell-mediated cytolysis and P-glycoprotein function in natural killer cells by verapamil isomers and cyclosporine A analogs. J. Clin. Immunol. 15:152–158.

Kobayashi, Y., Yamashiro, T., Nagatake, H., Yamamoto, T., Watanabe, N., Tanaka, H., Shigenobu, K. and Tsuruo, T. (1994). Expression and function of multidrug resistance P-glycoprotein in a cultured natural killer cell-rich population revealed by MRK16 monoclonal antibody and AHC-52. Biochem. Pharmacol. 48:1641-1646.

Maia, R.C., Silva, E.A.C., Harab, R.C., Lucena, M., Pires, V. & Rumjanek, V.M. (1996) Sensitivity of vincristine-sensitive K562 and vincristine-resistant K562-Lucena 1 cells to anthracyclines and reversal of multidrug resistance. Brazilian J. Med. Biol. Res. 29:467–472.

Maia, R.C., Wagner-Souza, K., Harab, R.C. & Rumjanek, V.M. (1996). Heparin reverses Rhodamine 123 extrusion by multidrug resistant cells. Cancer Letters 105, *in press.*

Neyfake, A.A. (1988). Use of fluorescent dyes as molecular probes for the study of multidrug resistance. Exp. Cell Res. 174:168–176.

Nooter, K. & Sonneveld, P. (1994). Clinical relevance of P-glycoprotein expression in haematological malignancies. Leukaemia Res. 18:233–243.

Pastan, I. & Gottesman, M. (1991). Multidrug resistance. Annu. Rev. Med. 42:277- 286.

Putney, J.W. (1990) Capacitative calcium entry revisited. Cell Calcium 11:611–624.

Riordan, J.R., Deuchars, K., Kartner, N., Alon, N., Trent, J. & Ling, V. (1985). Amplification of P-glycoprotein genes in multidrug-resistant mammalian cell lines. Nature 316:817–819.

Rumjanek, V.M., Lucena, M., Campos, M.M., Marques-Silva, V.M. & Maia, R.C. (1994). Multidrug resistance in leukemias: the problem and some approaches to its circumvention. Ciencia Cultura 46:63–69.

Sagara, Y., Fernandez-Belda, F., de Meis, L. & Inesi, G. (1992). Characterization of the inhibition of intracellular Ca^{2+} transport ATPases by thapsigargin. J. Biol. Chem. 267:12606–12613.

Thastrup, O., Cullen, P.J., Drobak, B.K., Hanley, M.R. & Dawson, A.P. (1990). Thapsigargin, a tumor promoter, discharges intracellular Ca2+ stores by specific inhibition of the endoplasmic reticulum Ca2+ ATPase. Proc. Natl. Acad. Sci. USA 87:2466–2470.

Thiebaut, F., Tsuruo. T., Hamada. H., Gottesman, M.M., Pastan, I. & Willingham, M.C. (1987). Cellular localization of the multidrug-resistance gene product in normal human tissues. Proc. Natl. Acad. Sci. USA 84:7735–7738.

Tsuruo, T., Iida, H., Ohkochi, E., Tsukagoshi, S. & Sakurai, Y. (1983). Establishment and properties of a vincristine-resistant human myelogenous leukemia K562. Jpn. J. Cancer Res. (Gann) 74:751–758.

Ueda, K., Cardarelli, C., Gottesman, M.M. & Pastan, Y. (1987). Expression of a full length cDNA for the human MDR1 gene confers resistance to colchicine, doxorubicine and vimblastine. Proc. Natl. Acad. Sci. USA 84:3004–3008.

Waldron, R.T., Short, A.D., Meadows, J.J., Gosh, T.K. & Gill, D.L. (1994). Endoplasmic reticulum calcium pump expression and control of cell growth. J. Biol. Chem. 269:11927–11033.

Willisch, A., Noller, A., Handgretinger, R., Weger, S., Nussler, V., Niethammer, D., Probst, H. & Gekeler, V. (1993). MDR 1/ P-glycoprotein expression in natural killer (NK) cells enriched from peripheral or umbilical cord blood. Cancer Letters 69:139–148.

11

REGULATION OF NEURONAL PROTEIN SYNTHESIS BY CALCIUM

J. R. Sotelo,[1] J. M. Verdes,[1,2] A. Kun,[1,3] J. C. Benech,[1,2]
J. R. A. Sotelo Silveira,[1,4] and A. Calliari[1,2]

[1]Laboratorio de Proteínas & Acidos Nucleicos del Sistema Nervioso
División Biofísica, Instituto de Investigaciones Biológicas Clemente Estable
Av. Italia 3318, Montevideo 11600, Uruguay
[2]Area Biofísica, Departamento de Biología Celular y Molecular
Instituto de Biociencias, Facultad de Veterinaria
Universidad de la República
Montevideo, Uruguay
[3]Unidad Asociada Biofísica
Instituto de Biología, Facultad de Ciencias
Universidad de la República
Montevideo, Uruguay
[4]Departamento de Biología Celular
Instituto de Biología, Facultad de Ciencias
Universidad de la República
Montevideo, Uruguay

1. INTRODUCTION

Neurons are an extreme example of a cell type whose architecture must have an influence on its local metabolism. Their extremely long projections, specially when axons are considered, together with ultrastructural characteristics such as the absence of concluding evidences for the presence of ribosomes inside axons, induced neurobiologists to propose different maintenance paradigms for the different parts of the neuron. Three neuronal territories will be discussed here, a) the axon; b) the soma and dendrites; and c) the synapse. The characteristics of the axonal territory will be described first. In accordance to the widespread dogma of the absence of ribosomes in axons, axonal proteins (structural ones or those functionally destinated to membranes or nerve endings) should be synthesized in the somatic territory and conveyed after towards the axon via two different velocity transport mechanisms [0.5 to 5 mm/day and 420 mm/day, respectively (for review see Ochs, 1982)]. The possibility that the axonal territory could synthesize some of its proteins has not been generally accepted. Thus, the half-life of each cytoskeletal axonal

Calcium and Cellular Metabolism: Transport and Regulation, edited by Sotelo and Benech.
Plenum Press, New York, 1997

protein should be estimated to be several times longer than the duration of their transport to the nerve endings. For instance, if we consider neurofilament proteins (Nf)—which normally are transported at a rate of 1mm/day—in a 1 meter length nerve, they will arrive to the nerve endings 1000 days after their synthesis. This could mean that if the half-life of Nf is 1000 days, 50% of the Nf transported to the nerve endings will be degraded before the original bulk arrives to its destination. Even with a half life of 4000 days (about 11 years), a 1/16 of the Nf transported to the nerve endings will not arrive given that they will be degraded. The above mentioned problem have been extensively discussed by Alvarez & Torrez (1985), whom proposed that under these conditions the most peripheral portions of axons should always be partially deprived of cytoskeletal proteins. Furthermore, it is well known from Cajal's studies (1928) that the total volume of the axonal territory in the longest axons is more than three order of magnitude greater than the somatic territory. On the other hand, the branching of axons at their ends will increase instead of diminish their total volume, increasing even more their possible hindrance. However, axons have been generally found to be homogeneously healthy throughout their length (see Alvarez & Torrez, 1985). Consequently, a local system of protein synthesis should be considered to maintain the homogeneity of axons; otherwise, some special mechanisms for protecting axonal cytoskeletal proteins from degradation must be found in axons. Thus, whether the axonal territory may synthesize at least a part of its own proteins or not, should be a main question, the answer of which—if positive—would change the point of view of a lot of normal or pathological neuronal functions which are related to protein synthesis.

Nevertheless, a set of experiments demonstrating the existence of local protein synthesis in the axonal territory will be shown here. It will also be shown here that local protein synthesis is regulated by calcium. Bostrom & Bostrom (1990) and Kimball & Jefferson (1992) have reviewed and shown in various cell types that calcium regulates eukaryotic protein synthesis, as well as other second messengers also do (Bostrom, et al. (1987). Little literature is related to the role of calcium on the regulation of eukaryotic protein synthesis in neurons. On the other hand, Ca^{2+} showed to be not only important in neurotransmission of excitable cells but also to participate as a second messenger in many other intracellular metabolic pathways. The demonstration that calcium would also regulate protein synthesis in neurons is extremely important, because it would ligate both: membrane electrical activity and protein synthesis. If Ca^{2+} regulates protein synthesis in neurons, it would be different whether we consider the soma, the dendrites, the axons or the nerve endings. For instance, the somatic protein synthesis may be regulated by calcium at two different levels (among others): a) the translation of the mRNAs at the ribosome level (see Bostrom & Bostrom, 1990); or b) the transcription processes in the nucleus throughout immediate early genes (Greenberg et al., 1986) or late response genes (reviewed by Williamson & Monck, 1989; Crabtree, 1989, Rao, 1991). On the other hand, the protein synthesis in the dendritic territory, in the axonal territory, or in the synaptic territory would be mostly regulated by calcium at the polyribosome level, since there is no DNA inside axons other than the mitochondrial DNA. Only if glial cells are considered as a possible source of mRNAs that codify for axonal proteins, the second possibility (the regulation of transcription by calcium), arises as a reliable option. In this regard as already mentioned, from now on, a set of experiments demonstrating the existence of *local eukaryotic protein synthesis in axons* will be described. Furthermore, a different set of experiments will be described in order to demonstrate that *calcium regulates eukaryotic protein synthesis in axons*. Finally, a set of experiments designed to characterize the *regulation by calcium of eukaryotic protein synthesis in dissociated cultured Dorsal Root Ganglion neurons* will be described.

2. LOCAL PROTEIN SYNTHESIS IN AXONS

2.1 Vertebrate Axons

The first electron microscope observations performed in order to examine the dynamics of changes occurring in the proximal stump of a sectioned peripheral nerve in mammals were done in the early sixties by Wettstein & Sotelo Sr. (1962). These observations raised a lot of important questions. The tip of the sectioned axons had plenty of newborn vesicles, probably provided from the smooth endoplasmic reticulum or from the axolemma. These vesicles appeared as soon as 30 min after sectioning the nerve. Some studies performed in our laboratory regarding the dynamics of orotic acid uptake (Benech, C. R. et al. 1968) into peripheral nerves and the rate of incorporation of radioactive aminoacids into the precipitable protein fraction of the normal and sectioned nerves suggested them that axons would be involved in the so-called local protein synthesis. The way in which the rate of uptake of radioactive aminoacids into the nerve changed after sectioning the nerve and the differences of protein synthesis between distal to proximal stumps in the course of time (Benech, C. R. unpublished results), also suggested them that axons were involved in this process. An early autoradiographic study made by Singer & Salpeter (1968), suggested the possibility that some proteins may be locally synthesized in the Schwann cell and transferred afterward to the axon. The presence of labelled proteins and RNA inside the axons was confirmed autoradiographically by Benech, C. R. et al. (1982), as well as labelled proteins inside the myelinated axons of the cat (Contreras et al., 1983) and the Mauthner axon of the goldfish (Alvarez & Benech, C. R., 1983). In the experiments carried out by Benech, C. R. et al. (1982), a pool of tritiated aminoacids or tritiated uridine were alternatively used for 30 min *in vivo* labelling of 5 mm of one of the proximal stumps. The autoradiograms of the uridine or aminoacids labelled proximal stumps showed silver grains over Schwann cells, myelin and axoplasm. Based on these results we proposed that: a) local protein synthesis exists in nerve fibers and it occurs far from the soma; b) it is at least partially independent from neuronal soma; c) the inhibition of protein synthesis by cycloheximide but its very low inhibition by chloramphenicol demonstrate that local protein synthesis is of the ribosomal eukaryotic type and not of the mitochondrial type, d) the RNA synthesis has the same cellular distribution; e) the effect of RNAse digestion and actinomycin D show that uridine is actually incorporated to RNA; f) silver grains found over those axons labelled with uridine were not associated to specific structures; g) if mitochondria are disregarded, newly synthesized RNA may only be originated in Schwann cells.

These results suggest that the newly synthesized RNA may be involved in the *de novo* synthesis of locally labelled proteins and the RNA may be synthesized in Schwann cells. The results reported here are in agreement with those from Koenig dealing with the synthesis of the enzyme acetylcholinesterase (1967, I, III) or the *in vitro* incorporation of tritiated precursors into proteins and RNA (1967, IV). Local protein synthesis was also studied by Tobias & Koenig (1975a, b) and by Frankel & Koenig (1977, 1978). The same authors detected the components of the protein synthesis machinery in the Mauthner giant axon of the goldfish (Koenig, 1979). It was also shown that axons in culture separated from the neuronal body are able to synthesize cytoskeletal proteins and that this synthesis is cycloheximide sensitive (Koenig & Adams, 1982; Koenig, 1989).

In other series of experiments of our laboratory (unpublished results), the sciatic nerve was crushed instead of cut and both stumps were incubated together with radioactive aminoacids. The washing of soluble radioactive aminoacids was performed simultaneously in both stumps before separating them. The proximal and distal stumps were separated just before Araldite embedding. Under these circumstances, autoradiographies of both stumps showed labelling in the axons, but the grain density was always higher in the axons of the proximal than in those of the distal stumps. Protein synthesis activity of axons of the proximal stump kept growing until it reached a plateau when arriving at the 8th day after crushing, but the grain density of distal stump axons followed an exponential decay. This means that axons of both stumps have different metabolic rates. These experiments may suggest that distal stump axons locally synthesize proteins during a limited period of time after disconnection. Thus, this synthesis should be dependent of the connection to the neuronal soma since the proximal stump can continue synthesizing proteins until the 8th day after injury.

These experiments raised a lot of questions, among which we emphasize the following: a) what kind of proteins are locally synthesized? We know that some of the newly synthesized proteins, as well as some of the newly synthesized RNAs, are inside the axon because we saw silver grains homogeneously distributed over the axoplasm with both specific precursors, but b) where are the proteins actually synthesized? c) what kind of RNA is present in the axon? d) are ribosomes actually present inside the axons?

2.2. Neurofilament Proteins Are Locally Synthesized

Neurofilaments are the specific intermediate filaments of neurons. In rats they are constituted by three protein subunits of 68 kDalton (kDa), 160 kDa and 200 kDa respectively. They are named as the low (L), medium (M) or high (H) molecular weight subunits. The 68L subunit is the constituent of the neurofilament core, while the other two are the constituents of the branches. Their two main characteristics are: a) they are neuron specific proteins, and b) they are considered as being slowly transported to the axon at a rate of 0.1 to 5 mm a day (for review see Ochs, 1982). These two main characteristics suggested us to study the possibility that these proteins may be locally synthesized. In this regard, the sciatic nerves of rats were processed as follows: 1) both sciatics of the same rat were sectioned and after 23 hours one of the proximal stumps was labelled *in vivo* with ^{35}S-methionine, as already described (Benech, C. R. et al., 1982, Sotelo et al., 1992), during one hour or four hours before excising them. Afterwards, they were desheathed and homogenized. 2) In another set of experiments, 4 cm of both sciatic nerves were excised, cut in 1 cm pieces, desheathed and labelled *in vitro* with ^{35}S-methionine and the effect of calcium movilization from nerves was tested (to be described later). The *in vivo* set of experiments was processed afterwards as follows: a) after homogenization of the nerves, samples were bidimensionally electrophoresed; b) blotted to nitrocellulose; c) neurofilament subunits were identified by immunostaining them using monoclonal comercial antibodies (Amersham); d) the nitrocellulose paper was autoradiographed (X-ray film), and e) finally the coincidence of both, the antibody spot (developed by alkaline phosphatase) and the autoradiographic spot were analyzed. There was a total coincidence of both spots, not only in shape, but in position coordinates. A one hour pulse experiment was enough to label the 68 kDa and 160 kDa, while four hours of incubation were needed to label the 200 kDa. These results clearly demonstrate that the three neurofilament subunits can be locally synthesized at least during the period of the early reaction to injury. The rate of the synthesis of the three subunits was higher *in vivo* than *in vitro*. These results were in agreement with our former finding that the distal stumps showed a lower rate of protein synthesis than the proximal stump. The *in vitro* experiments resemble the labelling of the distal stump.

2.3. Detection of Neuron-Specific mRNAs in Nerves

2.3.1. In Situ Hybridization. The 68 kDa (L) neurofilament subunit encoding mRNA was traced in the sciatic nerve of rats using a specially designed riboprobe (Lewis, & Cowan, 1985). After obtaining the above mentioned results showing that injury enhanced the synthesis of the neurofilament subunits, *in situ* hybridization (ISH) and RT-PCR (see below) were assayed in the proximal (ISH) or proximal and distal stumps (PCR) of the sectioned sciatic nerves. The nerves were fixed, frozen and sectioned in a cryostat. Cryostat sections obtained were incubated with an antisense riboprobe. This probe was labelled with digoxigenin-UTP by *in vitro* transcription of mouse 68kDa(L)-cDNA cloned in PBS(+/−). The probe was detected using an antibody against digoxigenin ligated to alkaline phosphatase and developed as usual. Control experiments were performed using the sense riboprobe (same labelling). Spinal cord cryostat sections were used as positive control tissues. Observed at the light microscope the reaction was found over motor neuron soma. Nerve sections showed reaction over Schwann cells as well as myelin, and despite some axons showed reaction over the axoplasm, the reaction was mainly observed in the frontier between the Schwann cell and the axoplasm. Obviously, to define exactly if the above-mentioned reaction is situated in the myelin or in the periphery of the axoplasm, the observation under the electron microscope is needed. The presence of the 68 kDa (L)-mRNA on the Schwann cells, myelin and axoplasm, suggested the idea that this mRNA may be locally synthesized in the Schwann cell and transferred thereafter to the axon or at least to the frontier between both cells. It has been found that glial cells are able to synthesize in vitro the 160 kDa neurofilament subunit (Kelly et al., 1992) and also the distal stump of the sectioned nerve (Roberson et al., 1992). The transport of poly-A RNA has been early reported by Bondy et al. (1977) and recently it has been found as being axonally transported in the hypothalamo-hypophyseal tract of rats (Mohr et al., 1991). The matter of the presence of RNA inside the axons has been studied for some time in different paradigms. For instance Edström et al. (1969) found RNA in the Mauthner axon, but it was Koenig (1979) who determined what RNA species were there. Shyne-Athwal et al. (1989), proposed that only tRNA is present in axons and it may be involved in the post-translational modification of already synthesized proteins in the axoplasm. The situation is different when invertebrate axons are considered. Some researchers claimed that in the squid giant axon only transfer RNA was present in the axoplasm (Black & Lasek, 1977). Consequently with this idea they reported evidences for the transfer of newly synthesized proteins from glia to the axoplasm of the squid giant axon, proposing a new hypothesis of the metabolic relationship between glia and axons at least in this paradigm (Gainer et al., 1977; Lasek et al., 1977; Tytell & Lasek, 1984). Completely different results were obtained in the squid giant axon by Giuditta et al. (1977) who found not only tRNA but elongation factors for protein synthesis, amynoacyl tRNA synthetase and also ribosomal RNA (Giuditta et al., 1980). Messenger RNA was not only detected in the squid giant axon (Giuditta et al., 1986), but its *in vitro* translation yielded a lot of neuron specific proteins. Here we will not review the work of this group extensively, because this will be done by Giuditta in Chapter 12, but we want to emphasize that they cloned more than one hundred mRNA obtained from the extruded axoplasm of the squid giant axon among which it was found β-actin, β-tubulin (Kaplan et al. 1992), Kinesin (Gioio et al., 1994) or enolase (Chun et al., 1995). On the other hand, early works of Giuditta et al. (1968) demonstrated the existence of protein synthesis in the isolated giant axon of the squid. Latelly active polysomes were detected in the axoplasm by Giuditta et al. (1991). Finally, in contrast with the hypothesis of transference of proteins from glia to the axon proposed by Lasek et al. (1977) and Gainer et al. (1977), Rapallino et al. (1988) detected the synthesis of axoplasmic RNA species in the glia surrounding the giant axon, that were transferred later to the axoplasm. Regarding our own above described results as well as the literature, we were prompted to search evidences for the presence of the other two neurofilament subunits mRNAs in the proximal and distal stumps of severed sciatic nerve of the rat.

2.3.2. Reverse Transcriptase-Polymerase Chain Reaction (RT-PCR). Specific primers were designed for each mRNA that codifies for the three neurofilament subunits. Total RNA was purified from the proximal and distal stumps (separatelly) of both sectioned sciatic nerves of more than ten rats. Poly-d(T) was used as a primer to synthesize cDNA. The cDNA obtained was amplified by PCR, using each couple of primers separately. The products of amplification were electrophoresed (polyacrylamide) and stained with silver nitrate. The proximal and distal stumps showed the presence of the expected size amplification product correspondent to the primers designed for the three mRNAs that codifies for the neurofilament subunits. The products of amplification were cloned and sequenced and the sequence corresponded to the three neurofilament transcripts. These results suggest that the three mRNAs are present in both nerve stumps. This is the first evidence reporting the presence of mRNAs that codifies for the three neurofilament subunits in the proximal stump of the sectioned sciatic nerves. Recently, using Northern blot, the transcripts for the 68 kDa subunit also was identified in the proximal stump. The latter together with the results of *in situ* hybridization and the autoradiographies of the immunoblots of the 3 neurofilament subunits, the results of RT-PCR strongly support the idea of the existence of an active protein synthesis machinery in the injured sciatic nerve that is independent of that of the neuronal soma.

2.4. Detection of Ribosomes in the Axonal Territory

The existence of ribosomes and polyribosomes inside the axons has been systematically disregarded despite they have been found in that place a limited number of times (Dimova & Markov, 1976; Palay et al., 1968; Peter et al., 1970; Tennyson, 1970; Skoff & Hamberger, 1974; Steward & Ribak, 1986; Pannese & Ledda, 1991). Two early studies made by Zelena (1970, 1972) who have intentionally searched for ribosomes in mammal axons by conventional electron microscopy described ribosomes in the proximal portions of sciatic axons and in axons passing throughout the dorsal root ganglion. The other mentioned findings were mostly done by chance when the authors were studying nerves with their minds oriented to other goals. On the other hand, Martin, Fritz & Giuditta (1989) visualized polyribosomes in the postsynaptic area of the squid giant synapse using phosphorous electron spectroscopic imaging (ESI). Crispino et al. (1996, submitted), also using ESI, detected ribosomes and polysomes in presynaptic nerve endings of the squid optic lobe and some axons of the squid optic nerve. In our laboratory we searched for ribosomes in vertebrate axons using a monoclonal antibody against S1 ribosomal protein, which was kindly supplied by B. Hügle (Hügle et al., 1985), using a peroxidase ligated secondary antibody. The analysis of the cryostat rat nerve sections at the light microscope after incubation with the monoclonal antibody, yielded a signal mainly localized in the border line between the Schwann cell and the axoplasm. A similar result was obtained in the Mauthner axons of fish. The latter results are in agreement with those reported by Koenig & Martin (1996) using the specific fluorescent nucleic acid dye named Yoyo-1. The above-mentioned dye selectively stained some cortical plaque-like structures of the extruded Mauthner axoplasm. The analysis of phosphorous content of this structures by

Electron Spectroscopic Images (ESI), showed a clear coincidence of ribosome and polysome-like particles with the cortical plaque-like structures detected by Yoyo-1 and ESI. On the other hand, a polyclonal antibody developped in rabbit against the purified ribosomes of the brain cortex of the rat, showed cross-reactivity (Immunocytochemistry Electron Microscopy) with ribosomes of the Rough Endoplasmic Reticulum (RER) of different cells of the squid, like glia cells or neurons. The antigen used to develop the polyclonal antibody was obtained by sucrose gradient ultracentrifugation. The ribosomal RNA content of the ribosomal fraction was tested by electrophoresis showing the clasical 18s and 28s ribosomal RNAs. The ribosomal fraction was examined at the electron microscope and the uranyl acetate negative staining showed the typical ribosomal profiles. The immunoblotting of the ribosomal fraction of the rat brain cortex fraction as well as the ribosomal fraction of the squid giant fiber lobe, showed in both a immunoreaction of at least 10 protein bands of similar molecular weights all below 60 kDalton. This latter result is in agreement with the idea that ribosomal proteins are highly preserved throughout evolution. When the polyclonal antibody was tested in the small as well as in the giant axons of the squid, it evidenced the presence of ribosomes and polyribosomes in the axoplasm of both kinds of fibers. Ribosomes and Polyribosomes were normally found in the peripheral part of the axons. They were found associated to the cytoskeleton or to some axoplasmic matrix not well defined yet (Sotelo et al., 1997, submitted).

3. REGULATION OF NEURON PROTEIN SYNTHESIS BY Ca^{2+}

As it has been extensively reviewed by Carafoli in the first introductory chapter of the present book, evolution has selected calcium to be one of the most important signalling cations. In this regard, the $[Ca^{2+}]_i$ (cytosolic) should be maintained extremelly low (nanomolar range). Thus, a transitory increment of the $[Ca^{2+}]_i$ can be sharply distinguished (from background noise) by the cell as a signal able to trigger some internal metabolic pathways. A large variety of Ca^{2+} sequestering proteins, cell membrane Ca^{2+} pump, endoplasmic reticulum Ca^{2+} pump, Na$^+$-Ca^{2+} exchanger are the described mechanisms through which calcium will be normally transported out of the cell or sequestered in internal stores in order to maintain a low cytosolic $[Ca^{2+}]$. On the other hand, Ca^{2+} channels would be the gates that should be transitorily opened in cell membranes to translate external stimulous, or in reticulum membranes to translate intracellular stimulous.

It has been largely shown in non-neuronal cultured cells that the entry of Ca^{2+} to the cytosol through the plasma membrane or released from the internal stores induces a strong inhibition of protein synthesis (Bostrom & Bostrom, 1990). There are several places where regulation of translation can be performed. For instance, the asociation of the mRNA to the ribosomal complex, a number of proteins associated with the translational apparatus including initiation factors, elongation factors and termination of translation (Bostrom et al., 1989; Chin et al., 1987; Fawell et al., 1989). Particularly, Elongation Factor 2 is reported to be phosphorilated and inhibited in reticulocytes lysates by calmodulin-dependent protein kinase III (Nairn & Palfrey, 1987; Ryazanov, 1987). Elevation of cytosolic free Ca^{2+} concentration at the light of these *in vitro* results would be in the basis of the inhibition of protein synthesis detected in intact reticulocytes exposed to the combination of Ca^{2+} and Ca^{2+} ionophores. Finally, gene expression can be regulated not only in the translational process, but also at the transcriptional level. As already mentioned, the nucleus has its own mechanisms of control of free $[Ca^{2+}]$ (Santella, L., personal communication) completely independent from the cytosolic free $[Ca^{2+}]$. The transitory change of the nucleus [Ca^{2+}] has been shown to induce the transcription of specific genes. The participation of calcium in neuronal plasticity has been extensively discussed (see Nelson & Douglas, 1994), throughout different mechanisms, among which was gene transcription. Nevertheless, the regulation of gene expression at the transcriptional level will not be the subject of the present chapter. Furthermore, it will be proposed here that if protein synthesis in neurons also occurs in axons and in the presynapsis (see Giuditta's chapter), it may also be regulated by calcium in the axonal and synaptic territory as well. From now on, we will describe a series of experiments showing how: a) local protein synthesis in the axonal territory may be regulated by calcium and later on a series of experiments showing: b) how somatic protein synthesis may be regulated by calcium. On the other hand, in the chapter of Benech et al., the regulation of protein synthesis by calcium in the synaptic territory using as a paradigm a synaptosomal preparation (Crispino et al., 1995) will be described.

3.1. Regulation of Local Nerve Protein Synthesis by Ca^{2+}

The sciatic nerves of rats were used *in vitro* to measure local protein synthesis. Desheathed 5 mm pieces of nerves were incubated at 37°C for 1 hour in Ringer-Tyrode solution containing ^{35}S-methionine plus 2 mM EGTA and either, one specific calcium ionophore, A23187 or Ionomycin. When the effect of ionomycin was assayed, different pieces of nerve were incubated changing the ionophore concentration from 0.5 μM to 10 μM. The same was done with A23187. In control experiments the incubation solutions contained the normal [Ca^{2+}] without EGTA. The TCA precipitable radioactivity (incorporation of methionine to newly synthesized proteins) measured in each piece of nerve was standardized to the control (normal Ca^{2+} concentration) pieces of nerve.

A similar dose-dependent inhibition of local protein synthesis was observed with both ionophores. Inhibition of protein synthesis arrived to 80% of control just between 1 to 2 μM ionophore concentration. The latter result strongly suggests that the inhibition of protein synthesis obtained in this condition was induced by the movement of calcium from

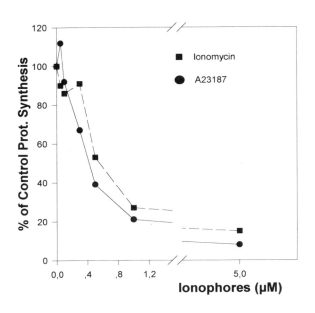

Figure 1. Effect of two specific Ca^{2+} ionophores (A23187 and ionomycin) on *in vitro* sciatic nerve protein synthesis. Protein synthesis of desheathed pieces of sciatic nerves was measured while the concentration of either A23187 (full circles), or ionomycin (full squares), varied in the μMolar range (μM). In ordinates: the percentage of control radioactive methionine incorporation to proteins of treated sciatic nerve was represented. A dosis dependent inhibition of protein synthesis was recorded.

the internal stores. In the presence of A23187 or Ionomycin the flux of calcium through-out plasma membrane and the endoplasmic reticulum membranes will be enhanced. In the meanwhile, EGTA chelates external Ca^{2+} enhancing the gradient from the internal stores to extracellular fluids. This situation finally results in a depletion of the content of Ca^{2+} of the internal stores. Our interpretation of the local protein synthesis inhibition obtained in this condition is similar to that of Bostrom & Bostrom (1990) who consider that a fundamental condition to have a normal rate of protein synthesis is maintaining a high calcium concentration in the internal store. Nevertheless, the relative importance of the fact that this condition also reduced drastically the concentration of Ca^{2+} of cytosol can not be disregarded. As it will be discussed later, we believe that an optimum cytosolic calcium concentration is needed to maintain a normal rate of protein synthesis.

Finally, newly synthesized labelled proteins under control conditions were identified by SDS-PAGE and compared to the newly synthesized labelled proteins in nerves incubated with EGTA and ionomycin. A general inhibition of protein synthesis has been found. One dimensional electrophoresis may not be enough to determine if there are changes of individual protein synthesis, but it is clear that local protein synthesis is also inhibited, since the newly synthesized neurofilament bands totally disappear when inhibition of protein synthesis was almost complete or were disminished during partial inhibition. On the other hand, SDS-PAGE of experiments in nerves carried out in the presence of calcium ionophores did not show an increment of low molecular weight bands (to be published elsewhere). The latter result suggests that the inhibition of protein synthesis observed is not related to the activation of calcium dependent proteases. These results suggest that axonal protein synthesis (like it occurs with eukaryotic protein synthesis) may be regulated by the changes of Ca^{2+} concentration. At the view of the latter sentence, we should keep in mind that the site of action of calcium on protein synthesis is direct or indirectly related to the ribosome complex or to the proteins associated with the process of eukaryotic protein synthesis (Bostrom et al., 1987; Bostrom et al., 1989). In this regard, it is important to emphasize that the finding of calcium regulation of axonal protein synthesis suggests the existence of an axonal eukaryotic ribosomal protein synthesis machinery.

3.2. Regulation of Protein Synthesis by [Ca²⁺] in the Neuronal Somatic Territory

The main goal for beginning the study of the regulation of protein synthesis by calcium in Dorsal Root Ganglion (DRG) neurons, was to compare how this process occurs in the neuronal soma of origin of at least part of the axons composing the sciatic nerve. Primary culture of adult dissociated DRG neurons was characterized metabolically. In this regard, DRG neurons were dissociated by collagenase digestion, followed by trypsin digestion carefully interrupted by a trypsin inhibitor and purified in a one step Percoll (Pharmacia) gradient centrifugation. About 3000 to 5000 DRG neurons were seeded over L-polylysine covered plastic Petri dishes or plastic 24 multiwell plates (the same number of cells per well). DRG neurons were fed with F-12 medium supplemented with 10% Fetal Bovine Serum (FBS). Protein synthesis was evaluated during the second day of culture, because at that time the DRG neuron cultures are already stabilized and they have an average of 70% of cell survival. On the other hand, 90% of these surviving neurons has already well-developped neurites (Sotelo et al., 1991). DRG neurons protein synthesis was analyzed by incubating with Serum Free Hank's Balanced Salt Solution (HBSS) containing ³⁵S-methionine (³⁵S-met). At the end of the pulse labelling, DRG neuron proteins were precipitated with 10% TCA, resuspended and radioactivity counted by Liquid Scintillation Counting (LSC). The rate of protein synthesis was evaluated as cpm/cell or cpm/μg of protein content (measured by the method of BIORAD). The rate of protein synthesis was found to be linear until about 4 hours of labelling. When newly synthesized proteins were analized by SDS-PAGE electrophoresis or bi-dimensional electrophoresis and fluorography, each labelled culture to be electrophoresed contained more than 20,000 neurons labelled in the same plastic Petri dish.

It is important to point out, that normally a living cell must maintain its cytosolic $[Ca^{2+}]$ at the nanoMolar range in order to survive. This should be important to the cell because it will permit to use the fast transitory changes of $[Ca^{2+}]$ to on-off different metabolic pathways (see Carafoli's chapter). According to Bostrom & Bostrom, (1990) and Kimball & Jefferson, (1992) the normal rate of protein synthesis in reticulocytes should be ligated not only to a low cytosolic concentration, but also to a high $[Ca^{2+}]$ inside the Rough Endoplasmic Reticulum (RER). Regarding the above-mentioned conditions, the following strategies were used in order to demonstrate the dependence of protein synthesis of DRG to $[Ca^{2+}]$:

a. the use of Ca^{2+} chelating molecules like EGTA, BAPTA or BAPTA-AM (after AM intracellular hydrolysis) to test lower intracellular $[Ca^{2+}]$ than normal;
b. permeabilization of the plasma membrane as well as the internal store membranes to the flux of calcium using calcium ionophores (A23187, ionomycin; Liu & Herman, 1978), inhibiting internal Ca^{2+} pumps with thapsigargin (Jackson et al., 1988; Wong et al., 1993);
c. Combinations of both of the above mentioned conditions.

3.2.1. Effect of the Depletion of Intracellular Ca^{2+} by EGTA, BAPTA, BAPTA-AM. It has been demonstrated in other cell types (Preston & Berlin, 1992), that the use of extracellular EGTA in excess of the normal extracellular $[Ca^{2+}]$ produces an internal depletion of Ca^{2+} that induces an inhibition of protein synthesis as well. In this regard, the following experiments were performed labelling newly synthesized proteins during 1 hour:

1. ^{35}S-met incubation was made together with 2mM EGTA concentration (extracellular $[Ca^{2+}]$=1mM);
2. DRG neurons were pre-incubated in 2 mM EGTA 1 hour before the incubation with ^{35}S-met. One hour incubation in ^{35}S-met was made together with 2 mM EGTA;
3. DRG neurons were pre-incubated in 2 mM EGTA 1 hour before the incubation with ^{35}S-met. Incubation in ^{35}S-met was made in 1 mM $[Ca^{2+}]$, without EGTA;
4. Pre-incubation was made during 1 hour (2 mM EGTA), but with 300μM concentration of Ca^{2+} in excess (2.3 mM); incubation was with or without calcium;
5. Pre-incubation in 2 mM EGTA was performed during the 48 hours after seeding. One hour incubation with ^{35}S-met at the end of the 48 hours. Neuron survival was controlled at the end of the 48 hours;
6. The 48 hours pre-incubation in 2 mM EGTA was performed together with 2.3 mM Ca^{2+}. Neuron survival was controlled at the end of the 48 hours and 1 hour incubation with ^{35}S-met was also made.

The latter series of experiments yielded the following results:

In agreement with the above mentioned experiments (Preston & Berlin, 1992) one hour incubation in EGTA was not enough to completely deplete intracellular calcium stores, so no inhibition of protein synthesis was observed. A different result was obtained when neurons were preincubated for 1 hour in EGTA and incubation with the radioactive precursor was accompanied with EGTA, with which a clear protein synthesis inhibition was recorded.

When EGTA pre-incubation was followed by the incubation with ^{35}S-met, but restoring extracellular calcium concentration without EGTA, the rate of protein synthesis was restored. This result is similar to the one reported by Bostrom et al. (1989).

When EGTA preincubation and ^{35}S-met incubation were done together with an extracellular [Ca^{2+}] in excess (2.3 mM) of the extracellular EGTA concentration (2.0 mM), protein synthesis was not inhibited, suggesting two different inferences: a) the protein synthesis inhibition induced by EGTA (extracellular Ca^{2+} free; pre-incubation and incubation with EGTA) is not induced by a toxic effect of EGTA, but by the abscence of calcium; b) it confirms that the depletion of Ca^{2+} internal stores are produced by the presence of extracellular EGTA, because when extracellular EGTA is saturated by external calcium, there is no protein synthesis inhibition. The pre-incubation of DRG neurons by 48 hours in EGTA induced a 50% cell death (confirmed by the disappearance of cells). Specific tests for controlling if the remaining cells were alive were not performed. Nevertheless, neurons still attached after 48 hours of EGTA incubation, were probably just dead or near death because they showed no protein synthesis after 1 hour of ^{35}S-met incubation. The above-mentioned abrupt increment of cell death percentage suggests that cell death would be a consequence—among other factors—of an inhibition of protein synthesis induced by the depletion of calcium internal stores. Furthermore, if the 48 hours EGTA incubation is accompanied by an excess of calcium concentration (2.3 mM Ca^{2+}; 2.0 mM EGTA), neuron survival returned to the normal level and no protein synthesis inhibition was observed (EGTA was saturated by calcium). It is important to emphasize that the incubation or preincubation with EGTA would produce pH changes of the medium that must be carefully buffered, because the changes of hydrogen ions concentration should induce changes in protein synthesis not well understood yet. In this regard, the same experiments were repeated adding BAPTA (a well-known chelating agent which does not affect pH) to the extracellular fluids, instead of EGTA.

The experiments in which BAPTA substituted EGTA to chelate calcium also induced protein synthesis inhibition. To confirm that the inhibition of protein synthesis by the pre-incubation with extracellular EGTA or BAPTA is induced by the depletion of the intracellular calcium concentration, DRG neurons were incubated with increasing concentrations of BAPTA-AM (μM range). Afterwards, newly synthesized proteins were radioactivelly labelled with ^{35}S-met. A protein synthesis inhibition correlated to the increase of BAPTA-AM concentration was observed. The latter result may suggest the idea that the inhibition of protein synthesis may be produced by the depletion of cytosolic [Ca^{2+}], as well as a lowering of the [Ca^{2+}] of internal stores. BAPTA-AM entered into the DRG neurons, the BAPTA acetoxymethyl ester is hydrolized by internal sterases, then remaining inside the cell. Thus, the cytosolic [Ca^{2+}] will lower afterwards, as it was measured by Preston & Berlin (1992). On the other hand, considering that the concentration of internal calcium stores is normally in the mM range, it may be supposed that the protein inhibition recorded in this case could mainly be related to the depletion of cytosolic [Ca^{2+}], specially if intracellular BAPTA concentration (5 μM) should be three order of magnitude lower than the internal store [Ca^{2+}].

Finally, the above described experiments together with the experiments that will follow, suggest that an optimal cytosolic [Ca^{2+}] is needed for maintaining the normal rate of protein synthesis in cultured DRG neurons.

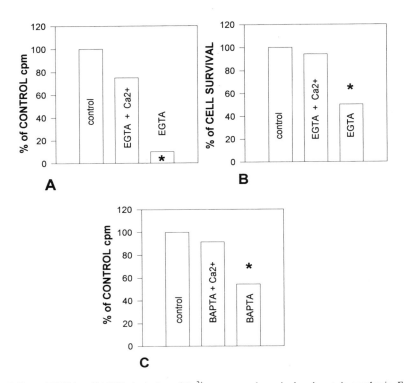

Figure 2. Effect of EGTA or BAPTA depletion of Ca^{2+} on neuronal survival and protein synthesis. Fig. 2A shows the effect of 48 hs EGTA pre-incubation on protein synthesis (standardized to control). Inhibition of protein synthesis is almost total. Fig. 2B shows the effect of 48 hs EGTA pre-incubation on neuron survival (evaluated just by counting cells still attached to the substrate), neuron survival changed from 80% in control neurons to 50% in EGTA preincubated neurons. When 48 hs EGTA pre-incubation was performed saturating EGTA with an excess of calcium (EGTA + Ca^{2+}, bar) neither protein synthesis nor neuron survival showed significative statistical differences when compared to control neurons (normal calcium no EGTA). A similar result is shown in the experiment represented in Fig. 2C where 48 hs EGTA pre-incubation is substituted by 24 hs pre-incubation in BAPTA. An inhibition of protein synthesis as compared to control is recorded (Bapta, bar). No inhibition is recorded when BAPTA is saturated with an excess of calcium (BAPTA + Ca^{2+}, bar). (*values statistically different to the correspondent control experiments).

3.2.2. Increasing the Cytosolic [Ca^{2+}]. Effect on Protein Synthesis. Two specific Ca^{2+} ionophores (A23187 and ionomycin) were used in order to increase the cytosolic [Ca^{2+}]. As it can be seen in Fig. 3, a strong inhibition of protein synthesis was associated to the increment of the concentration of both ionophores in the extracellular solution when the concentration of extracellular Ca^{2+} was 1 mM. These results strongly suggest that both ionophores are permeating the plasma membranes as well as the intracellular membranes of the Ca^{2+} internal stores. In this regard, the ionophores are equalizing [Ca^{2+}] in the three compartments. Since normally, the cytosolic [Ca^{2+}] is about the nM level, the introduction of an ionophore will equalize the cytosolic and the internal stores [Ca^{2+}] to the extracellular [Ca^{2+}] (1 mM). We wish to stress the fact that the extracellular volume must be considered as being infinite when compared to the other two above-mentioned compartments. These results are in agreement with those largely reviewed by Bostrom & Bostrom, (1990). This inhibition reached to more than 50% when the concentration of the ionophore was just 1 µM. When the same experiment was performed in the presence of 1 mM EGTA (varying ionophore

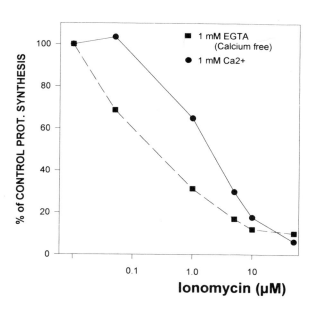

Figure 3. Effect of the Ca²⁺ specific ionophore (ionomycin) on neuronal protein synthesis. Cultured neurons were fed with solutions with increasing concentrations of ionomycin (μM) in the presence of calcium (1 mM [Ca²⁺]ᵢ, full circles), or in free calcium solution plus 1 mM EGTA (full squares). Simultaneously, the rate of protein synthesis of neurons treated in this way was standardized to control neurons. In ordinates: the percentage of control radioactive methionine incorporation to proteins of treated neurons is represented. A dosis dependent inhibition of protein synthesis was recorded.

concentration, and the extracellular solution without added calcium), an enhancement of the inhibition was observed. The latter result, already mentioned before, may be interpreted in the following way total Ca²⁺ depletion of both compartments, the cytosol and the internal stores, induced a strong inhibition of ribosomal protein synthesis. This inhibition was established faster than the inhibition obtained by the presence of the ionophores without EGTA. Thus, protein synthesis should be inhibited by the depletion of calcium, as well as the rising of the cytosolic calcium, suggesting that an optimal concentration of calcium should be needed to maintain the normal rate of protein synthesis. Finally, it is important to point out that the [Ca²⁺] of the internal stores must be in the millimolar range (Bostrom & Bostrom, 1990; Kimball & Jefferson, 1992).

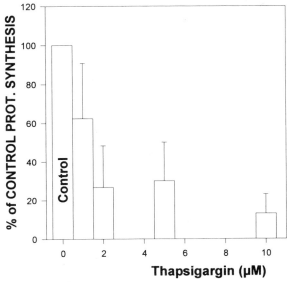

Figure 4. Effect of the SERCA family Ca²⁺ATPase inhibitor Thapsigargin on neuronal protein synthesis. Thapsigargin was offered to cultured neurons varying its concentration in the μM range (abscissa). Neuronal protein synthesis is evaluated and standardized to control neuronal protein synthesis (ordinates). A dosis dependent inhibition of protein synthesis was recorded.

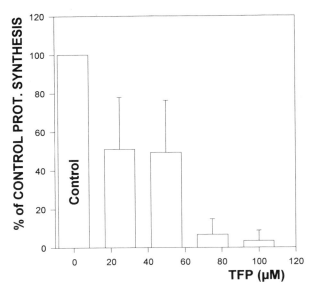

Figure 5. Effect of trifluoperazine (TFP) on neuronal protein synthesis. TFP was offered to cultured neurons varying its concentration in the μM range (abscissa). Neuronal protein synthesis was evaluated and standardized to control neuronal protein synthesis (ordinates). A dosis dependent inhibition of protein synthesis was recorded.

As already mentioned, the cytosolic $[Ca^{2+}]$ may rise by the inward of calcium through the plasma membrane (opening of the Ca^{2+} channels, inhibition of plasma membrane $Ca^{2+}ATPase$, etc.), or by the opening of the internal Ca^{2+} channels, or the inhibition of the internal $Ca^{2+}ATPase$. The effect of thapsigargin (a strong specific inhibitor of SERCA family $Ca^{2+}ATPases$; Lytton et al., 1991; Sagara & Inesi, 1991) over protein synthesis was tested here. As it can be seen in Fig. 4 there is a dosis dependent inhibition of protein synthesis by thapsigargin.

When thapsigargin inhibits $Ca^{2+}ATPase$, it stops the storage of calcium into the internal stores. This may inhibit protein synthesis in two different ways that cannot be determined by the experiments described here: a) an increament of the cytosolic $[Ca^{2+}]$ or a depletion of the internal store $[Ca^{2+}]$.

Finally, the use of a Calmodulin inhibitor like trifluoperazine (TFP) also produces an inhibition of protein synthesis. This result may support the already mentioned reports (Bostrom et al., 1989) suggesting that the phosphorilation of elongation factor 2 mediated by calcium through calmodulin may also be occurrying in DRG neurons. On the other hand, the inhibition of protein synthesis by TFP is also supporting the idea that a normal protein synthesis rate should require not only a high RER calcium concentration, but an optimum cytosolic calcium concentration. The latter may be so, because if calmodulin is involved in the regulation of protein synthesis, the changes of the cytosolic calcium concentration should also be involved.

4. CONCLUSIONS

The main goal of the present chapter is to demonstrate the importance of calcium in neurons as a second messenger that can be relating plasma membrane functions (as excitability) with cellular metabolism such as protein synthesis. The findings reported here clearly suggest that calcium should be important to regulate protein synthesis in the somatic territory. This means that protein synthesis in neurons may be modulated by the input of information from the environment or by the internal information of its own metabolism. On the other hand, the inhibition of local nerve protein synthesis together with the findings reported in the chapter of Benech et al. related to the regulation of protein synthesis in nerve endings of the squid suggest the possibility that local protein synthesis is regulated by calcium as well as somatic protein synthesis also is. They also suggest that the regulation of neuronal protein synthesis by calcium would be important to modulate all the processes that are changing the morphology of the neuronal soma or neuronal prolongations such as: regeneration and neuronal or synaptic plasticity. This modulation may also contribute to consolidate or transform short-term changes (that may be associated to changes in the environment) into long-term processes.

5. SUMMARY

The regulation of neuronal protein synthesis by calcium in the somatic and axonal territories of neurons is analyzed in the present chapter. A series of results showing the existence of a local protein synthesis in the axonal territory is demonstrated using different techniques: optical micro autoradiography; 1 or 2 dimensional electrophoresis and fluorography; immunoblotting with specific monoclonal antibodies against neurofilament proteins. On the other hand, the existence of the protein synthesis machinery responsible for the above-mentioned local protein synthesis in the axonal territory is also analyzed. In this regard, the presence of the mRNAs codifying for the neurofilament subunit proteins is also demonstrated using *in situ* hybridization; and the Reverse Transcriptase Polymerase Chain Reaction (RT-PCR) method. The presence of ribosomes and polysomes inside the squid giant axon is also demonstrated using a polyclonal antibody against ribosomes. Finally, a series of experiments showing the dependence of protein synthesis of the $[Ca^{2+}]_i$ in the two above mentioned territories are reported here. The use of specific calcium ionophores (like ionomycin and A23187), Ca^{2+} chelating agents (like EGTA, BAPTA and BAPTA-AM), thapsigargin (specific inhibitor of SERCA ATPases) and Trifluoperazine (TFP, inhibitor of calmodulin), while the rate of protein synthesis of dissociated cultured neurons is analyzed, shows that an optimum calcium concentration in the cytosol is needed in order to maintain the normal rate of protein synthesis. The demonstration of calcium regulation of protein synthesis in both neuronal territories is important because it relates membrane activity to gene translation. This regulation should be important to understand regeneration and plasticity in both territories.

6. ACKNOWLEDGMENTS

The present work was supported by the Grant CI1*93-0037 of the European Union, the Japanese International Cooperation Agency (JICA), BID-CONICYT, PEDECIBA, MEYC, Facultad de Ciencias and Facultad de Veterinaria (Universidad de la República). We wish to specially acknowledge the help of Mrs. A. Wittenberger on carefully reading the present manuscript.

7. REFERENCES

Alvarez, J., & Benech, C. R. (1983). Axoplasmic incorporation of Amino Acids in a Myelinated Fiber Exceeds That of Its Soma: A Radioautographic Study. Exp. Neurol. 82: 25–42.
Alvarez, J., & Torres, J. C. (1985) Slow axoplasmic transport: a fiction? J. Theor. Biol. 112:627–651.

Benech, C. R., Saá, E. A., & Franchi, C. M. (1968). In Vivo Local Uptake of C-14 Orotic Acid by Peripheral Nerve. Exp. Neurol. 22: 436–443.

Benech, C., Sotelo, J. R., Menéndez, J., & Correa-Luna, R. (1982). Autoradiographic Study of RNA and Protein Synthesis in Sectioned Peripheral Nerves. Exp. Neurol. 76: 72–82.

Black, M. M., & Lasek, R. J. (1977), The presence of transfer RNA in the axoplasm of the squid giant axon. J. Neurobiol. 8: 229–237.

Bondy, S. C., Purdy, J. L. & Babitch, R. (1977). Axoplasmic transport of RNA containing a polyadenilic acid segment. Neurochem. Res. 2:407–415.

Brostrom, C. O., & Brostrom, M. A. (1990). Calcium-dependent regulation of protein synthesis in intact mammalian cells; Ann. Rev. Physiol. 52: 577–90.

Brostrom, C. O., Chin, K. V., Wong, W. L., Cade, C., & Brostrom, M. A. (1989). Inhibition of traslational initiation in eukaryotic cells by calcium ionophore. J. Biol. Chem. 254: 1644–49.

Brostrom, M. A., Chin, K. V., Cade, C., Gmitter, D., & Brostrom, C. O. (1987). Stimulation of protein synthesis in pituitary cells by phorbol esters and cyclic AMP. Evidence for rapid induction of a component of traslational intiation. J. Biol. Chem. 262: 16515–23.

Chin, K., Cade, C., Brostrom, C. O., Galuska, E. M., & Brostrom, M. A. (1987). Calcium- dependent regulation of protein synthesis at traslational initiation in eukariotic cells. J. Biol. Chem. 262: 16509–14.

Chun, J. T., Gioio, A. E., Crispino, M., Giuditta, A., & Kaplan, B. B. (1995). Characterization of squid enolase mRNA: sequence analysis, tissue distribution, and axonal localization. Neurochem. Res. 20923–930.

Contreras, G., Carrasco, O. C., & Alvarez, J. (1983). Axoplasmic incorporation of aminoacids in myelinated fibers of the cat. Exp. Neurol. 82: 581–593.

Crispino, M., Perrone-Capano, C., Kaplan, B. B., & Giuditta, A. (1995). Neurofilament proteins are synthesized in nerve endings from squid brain. J. Neurochem. 61: 1144–1146.

Dimova, R. N., & Markov, D. V. (1976). Changes in the mitochondria in the initial part of the axon during regeneration. Acta Neuropathol. (Berlin) 36: 235–242.

Edstrom, A., Edstrom, J. E., & Hokfelt, T. (1969). Sedimentation analysis of ribonucleic acid extracted from isolated Mauthner nerve fibre components. J. Neurochem. 16: 53–66.

Fawell, E. H., Boyer, I. J., & Brostrom, C. O. (1989). A novel calcium-dependent phosphorylation of a ribosomal-associated protein. J. Biol. Chem. 254: 1650–55.

Frankel, R. D., & Koenig, E. (1977). Identification of major indigenous protein components in mammalian axons and locally synthesized axonal protein in hypoglossal nerve. Exp. Neurol. 57: 282–295.

Frankel, R. D., & Koenig, E. (1978). Identification of locally synthesized proteins in proximal stumps axons of the neurotomized hypoglossal nerve. Brain Res. 141: 67–76.

Gainer, H., Tasaki, I., & Lasek, R. J. (1977). Evidence for the glia-neuron protein transfer hypothesis from intracellular perfusion studies of squid giant axons. J. Cell Biol. 74: 524–530.

Gioio, A. E., Chun, J. T., Crispino, M., Perrone-Capano, C., Giuditta, A., & Kaplan, B. B. (1994) Kinesin mRNA is present in the squid giant axon. J. Neurochem. 63: 13–18.

Giuditta, A., Dettbarn, W. D., & Brzin, M. (1968). Protein synthesis in the isolated giant axon of the squid. Proc. Natl. Acad. Sci. USA. 59: 1284–1287.

Giuditta, A., Metafora, S., Felsani, A., & Del Rio, A. (1977). Factors for protein synthesis in the axoplasm of squid giant axon. J. Neurochem. 28: 1393–1395.

Giuditta, A., Cupello, A., & Lazzarini, G. (1980). Ribosomal RNA in the axoplasm of the squid giant axon. J. Neurochem. 34: 1757–1760.

Giuditta, A., Hunt, T., & Santella, L. (1983). Messenger RNA in squid axoplasm. Biol. Bull. 165: 526.

Giuditta, A., Hunt, T., & Santella, L. (1986). Messenger RNA in squid axoplasm. Neurochem. Intern. 8: 435–442.

Giuditta, A., Menichini, E., Perrone-Capano, C., Langella, M. Martin, R., Castigli, E., & Kaplan., B.B. (1991) Active polysomes in the axoplasm of the squid giant axon. J. Neurosci. Res. 28: 18–28.

Hügle, B., Hazan, R., Scheer, U., & Franke, W., (1985). Localization of ribosomal protein S1 in the granular component of the interphase nucleolus and its distribution during mitosis. J. Cell Biol. 100: 873–886.

Jackson, T. R., Patterson, S. I., Thastrup, O., & Hanley, M. R. (1988). A novel tumor promoter, thapsigargin, transiently increases cytoplasmic free Ca^{2+} without generation of inositol phosphates in NG115–401L neuronal cells; Biochem. J. 253: 81–86.

Kaplan, B. B., Gioio, A. E., Perrone-Capano, C. Crispino, M., & Giuditta, A. (1992) Beta-Actin and Beta-Tubulin are components of a heterogeneous mRNA population present in the squid giant axons. Mol. Cell Neurosci. 3: 133–144.

Kelly, B. M., Gillespie, C. S., Sherman, D. L., & Brophy, P.J., (1992). Schwann cells of the myelin-forming phenotype express neurofilament protein NF-M. J. Cell Biol. 118:397- 410.

Koenig, E., (1967). Synthetic mechanisms in the axon. I. Local axonal synthesis of acetylcholinesterase. J. Neurochem. 12:343–355.

Koenig, E., (1967). Synthetic mechanisms in the axon. III Stimulation of acetylcholinesterase synthesis by acti-nomycin-D in the hypoglossal nerve. J. Neurochem. 14: 429–435.

Koenig, E. (1967). Synthetic mechanisms in the axon. IV. *In vitro* incorporation of [^{3}H]precursors into axonal pro-tein and RNA. J. Neurochem. 14: 437–446.

Koenig, E. (1979). Ribosomal RNA in the Mauthner axon: implications for a protein synthesis machinery in the myelinated axons. Brain Res. 174: 95–107.

Koenig, E., & Adams, P. (1982). Local protein synthesis activity in axonal fields regenerating *in vitro*. J. Neuro-chem. 39: 386–400.

Koenig, E. (1989). Cycloheximide-sensitive [^{35}S]methionine labeling of proteins in goldfish retinal ganglion cell axons in vitro. Brain Res. 481: 119–123.

Koenig, E., & Martin, R., (1996). Cortical plaque-like structure identify ribosome-containing domains in the Mauthner cell axon. J. Neurosci. 16: 1400–1411.

Kimball, S. R., & Jefferson, L. S. (1992). Regulation of protein synthesis by modulation of intracellular calcium in rat liver; Am. J. Physiol., 263 (Endocrinol. Metab. 26): E958–E964.

Lasek, R. J., Gainer, H., & Barker, J. L. (1977). Cell to cell transfer of glial proteins of the squid giant axon. The glia-neuron protein transfer hypothesis. J. Cell Biol. 74: 501–523.

Lewis, S. A., & Cowan, N. J., (1985). Genetics, evolution, and expression of the 68,000-mol-wt neurofilament protein: isolation of cloned cDNA probe. J. Cell Biol. 100:843–850.

Lytton, J., Westling, M., & Hanley, M. R. (1991). Thapsigargin inhibits the sarcoplasmic or endoplasmic reticulum Ca^{2+}ATPase family of calcium pumps. J. Biol. Chem. 266: 17067–17071.

Liu, C., & Herman, T. E. (1978). Characterization of ionomycin as a calcium ionophore. J. Biol. Chem. 253: 5892–94.

Martin, R., Fritz, W., & Giuditta, A. (1989). Visualization of polyribosomes in the postsynaptic area of the squid giant synapse by electron spectroscopic imaging. J. Neurocytol. 18: 11–18.

Mohr, E., Fehr, S., & Richter, D. (1991). Axonal transport of neuropeptide encoding mRNAs within hypothalamo-hypophyseal tract of rats. EMBO J. 10: 2419–2424.

Nairn, A. C., & Palfrey, H. C. (1987). Identification of the major Nr 100,000 substrate for calmodulin-dependent protein kinase III in mammalian cells as elongation factor-2. J. Biochem. 262: 17299–17303.

Nelson, P. G., & Fields R. D. (1994). Calcium and Neuronal Plasticity. J. of Neurobiology 25: 219.

Ochs, S. (1982). Axoplasmic transport and its relation to other nerve functions. John Wiley & Sons, Inc. New York, USA.

Palay, S. L., Sotelo, C., Peters, A., & Orkand, P. M., (1968) The axon hillock and the initial segment. J. Cell Biol. 38: 193–201.

Pannese, E., & Ledda, M., (1991). Ribosomes in myelinated axons of the rabbit spinal ganglion neurons. J. Submi-crosc. Cytol. Pathol. 23: 33–38.

Peter, J. A., Palay, S. L., & Webster, H. De F. (1970) The fine structure of the nervous system. N.Y.: Harper and Row, p. 198.

Preston, S. F., & Berlin, R. D. (1992). An intracellular calcium store regulates protein synthesis in HeLa cells, but it is not the hormone-sensitive store. Cell Calcium 13: 303–312.

Ramon y Cajal, S., Studies on degeneration and regeneration of the nervous system, (1928). Vol. I, Translated by R. M. May. Oxford University Press, Oxford, reprinted Hafner, N. Y., 1968, p. 290.

Rapallino, M. V., Cupello, A., & Giuditta, A. (1988). Axoplasmic RNA species synthesized in the isolated squid giant axon. Neurochem. Res. 13: 625–631.

Roberson, M. D., Toews, A. D., Goodrum, J. F., & Morell, P., (1992). Neurofilament and Tubulin mRNA expres-sion in Schwann cells. J. Neurosci. 33: 156–162.

Ryazanov, A. G. (1987). Calcium /calmodulin dependent phosphorilation of elongation factor 2. FEBS Letts. 214: 331–334.

Sagara, Y., & Inesi, G. (1991). Inhibition of the sarcoplasmic reticulum Ca^{2+} transport ATPase by thapsigargin at subnanomolar range concentrations. J. Biol. Chem. 266: 13503–13506.

Shyne-Athwal, S. Chakraborty, G., Gage, E., & Ingoglia, N. A. (1989). Comparison of post-translational protein modification by amino acid addition after crush injury to sciatic and optic nerves of rats. Exp. Neurol. 99: 281–295.

Singer, M., & Salpeter. (1966). The transport of [^{3}H]L-histidine through the Schwann and myelin sheath into the the axon, including a reevaluation of myelin function. J. Morphol. 120: 281–293.

Skoff, R. P., & Hamberger, V., (1974). Fine structure of dendritic and axonal growth cones in embryonic chick spi-nal cord. J. Comp. Neurol. 153: 107–148.

Sotelo, J. R., Benech, C. R., & Kun, A. (1992). Local radiolabeling of the 68kDa neurofilament protein in rat sci-atic nerves. Neurosc. Letts. 144: 174–76.

Sotelo, J. R., Horie, H. Ito, S., Benech, C., Sango, K., & Takenaka, T. (1991). An in vitro model to study diabetic neuropathy. Neurosc. Letts. 129: 91–94.

Steward, O., & Riback, C. E., (1986) Polyribosomes associated with synaptic specialization on axon initial segments. J. Neurosci. 6: 3079–3085.

Tennyson, V. M., (1970). The fine structure of the axon and growth cone of the dorsal root neuroblast of the rabbit embryo. J. Cell Biol. 44: 62–79.

Tobias, G. S., & Koenig, E. (1975). Axonal protein synthesizing activity during the early outgrowth period followin neurotomy. Exp. Neurol. 49: 221–234.

Tobias, G. S., & Koenig, E. (1975). Influence of nerve cell body and neurilemma cells on local axonal protein synthesis following neurotomy. Exp. Neurol. 49: 235–245.

Tytell, M., & Lasek, R. J., (1984). Glial polypeptides transferred into the squid giant axon. Brain Res. 324: 223–232.

Wettstein, R., & Sotelo Sr., J. R.. (1963). Electron microscope study of the regenerative process of peripheral nerves of mice. Z. Zellforsch. 59: 708–730.

Wong, W. L., Brostrom, M. A., Kuznetsov, G., Gmitter-Yellen, & Brostrom, C. O. (1993). Inhibition of protein synthesis and early protein processing by thapsigargin in cultured cells. Biochem. J. 289: 71–79.

Zelena, J. (1970). Ribosome-like particles in myelinated axons of the rat. Brain Res. 24: 359–363.

Zelena, J. (1972). Ribosomes in myelinated axons of dorsal root ganglia. Z. Zellforsch. 124: 217–229.

12

LOCAL PROTEIN SYNTHESIS IN THE SQUID GIANT AXON AND PRESYNAPTIC TERMINALS

Antonio Giuditta, Mariana Crispino, and Carla Perrone Capano

Dipartimento di Fisiologia Generale e Ambientale
Università di Napoli "Federico II"
Via Mezzocannone 8
Napoli 80134, Italy

1. INTRODUCTION

The existence of a local system of axonal protein synthesis was proposed approximately 35 years ago to explain the similar rates of reappearance of acetylcholinesterase activity in proximal and distal regions of the cat hypoglossal nerve following irreversible inactivation of the enzyme (Koenig and Koelle, 1960). Since then this question has been approached with more direct techniques, by measuring the incorporation of radiolabelled aminoacids into the proteins of the axonal compartment following its separation from the closely apposed glial cells by microdissection or by autoradiographic methods (Koenig, 1984; Giuditta et al., 1990). The separation of axonal and glial compartments is more readily accomplished in large axons, such as the squid giant axon (Giuditta et al., 1968; 1983; 1991) and the Mauthner axon of the goldfish (Koenig, 1979; 1991; Koenig and Martin, 1996).

The data reviewed in this article demonstrate that the isolated squid giant axon (i.e., separated from its nerve cell bodies) is capable of extramitochondrial protein synthesis, and that active eukaryotic polysomes occur in the axon itself. Eukaryotic protein synthesis has likewise been detected in the large presynaptic endings of the squid photoreceptor neurons. The paper will also be concerned with related problems, such as the identification of the proteins synthesized by axoplasmic polysomes, and the cellular origin of axoplasmic RNA.

The giant axon of the squid utilized in our studies is localized in the most medial stellate nerve which innervates the distal regions of the mantle. In addition to the giant axon, each stellate nerve contains several axons of normal size which may be dissected away from the giant axon. Using a tiny roller, the axoplasm of the giant axon may be squeezed out from the remaining sheath, yielding about 5 mg of pure axoplasm.

Calcium and Cellular Metabolism: Transport and Regulation, edited by Sotelo and Benech.
Plenum Press, New York, 1997

2. PROTEIN SYNTHESIS IN THE SQUID GIANT AXON

2.1. Identification of an Axonal System of Eukaryotic Polysomes

The procedure just described was used to demonstrate the local synthesis of axoplasmic proteins in isolated giant axons incubated with radioactive aminoacids. The incorporation reaction was strongly inhibited by puromycin or cycloheximide (Giuditta et al., 1968). This initial observation was followed by a series of experiments aimed at the localization of the system of axonal protein synthesis. Two main possibilities were envisaged, according to whether the involved eukaryotic polysomes were thought to be localized in the axon itself or in the periaxonal glial cells. The latter alternative (glia-neuron protein transfer hypothesis) was proposed on the basis of a number of observations made on isolated or perfused giant axons (Lasek et al., 1974; 1977; Gainer et al., 1977). The hypothesis was grounded on the assumption that the axon lacked eukaryotic polysomes, and that the axonal proteins synthesized by the isolated giant axon originated in the periaxonal glial cells. Conversely, our hypothesis postulated the occurrence of active eukaryotic polysomes in the giant axon, notably in its cortical layers (Giuditta, 1980), but did not exclude the transfer of glial proteins to the axon. It may be noted that neither hypothesis challenged the existence of a local system of synthesis of axonal proteins, which was generally assumed to complement the set of proteins delivered to the axon by the nerve cell bodies.

In the first round of experiments examining the validity of our hypothesis, the axoplasm extruded from the giant axon was analyzed to ascertain the presence of all the soluble proteins and RNA factors required by the eukaryotic system of protein synthesis. This was proven by an experiment in which the negligible protein synthetic activity of purified rabbit reticulocyte polysomes was dramatically enhanced by addition of the post-mitochondrial fraction of the axoplasm (Giuditta et al., 1977). More detailed assays indicated that the axoplasm also contains elongation factors, tRNAs and aminoacyl-tRNA synthetases (Giuditta et al., 1977). Later electrophoretic analyses of purified axoplasmic RNA demonstrated that, in addition to a major 4S component, the axoplasm contains the two ribosomal RNAs (accounting for about 10% of the total RNA) and an additional RNA component slightly larger than 4S RNA (presumably a 7S RNA; Giuditta et al., 1980). These findings considerably extended previous electrophoretic data suggesting the presence of only the 4S RNA in the giant axon's axoplasm (Lasek et al., 1973). The identification of ribosomal RNA in the axoplasm supported the view that a eukaryotic system of protein synthesis is present in the axon (Giuditta, 1980). This possibility was further confirmed by the demonstration that the axoplasmic 4S RNA contains several discrete tRNAs identified by specifically charging them with single radioactive aminoacids (Ingoglia et al., 1983). Individual aminoacyl-tRNA synthetases were also detected in the axoplasm (Ingoglia et al., 1983).

A turning point in our studies was reached with the demonstration that the axoplasm contains a whole family of mRNAs (Giuditta et al., 1983; 1986). This result was obtained in cell-free translation analyses of purified axoplasmic RNA using a rabbit reticulocyte lysate supplemented with [^{35}S]methionine. As shown in Figure 1, approximately 50 radioactive protein bands could be discerned in the translation products of axoplasmic RNA. Some of the major bands had electrophoretic mobilities corresponding to those of the main cytoskeletal proteins, such as α- and β-tubulin, β-actin, and the neurofilament proteins. While many axoplasmic protein bands corresponded to translation products of the RNA purified from the parent cell bodies of the giant axon (giant fibre lobe), a few components (notably, a 115 kDa protein) were lacking in the translation pattern of the nerve cell bodies. This provided the initial indication that some mRNAs are selectively localized

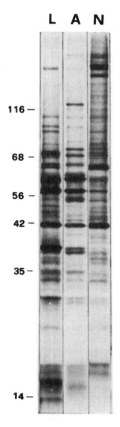

Figure 1. Electrophoretic patterns of the translation products of mRNAs purified from the giant fibre lobe (L), the axoplasm of the giant axon (A), and the stellate nerve (N). Molecular weight markers are shown on the left lane. Cell-free translation analyses were made using a rabbit reticulocyte lysate.

in the axon. Conversely, some translation products of the cell bodies RNA were not present in the axoplasmic translation pattern. To our knowledge, these data represent the first demonstration of a whole population of mRNAs in a mature native axon, and of the particular composition of the family of axonal mRNAs in comparison with the perikaryon.

The possibility that the above results were due to glial contamination occurring during the extrusion step was excluded by an experiment in which RNA was separately extracted from a sample of extruded axoplasm and from the corresponding sample of axon sheaths. The translation patterns clearly indicated that the axoplasmic bands were much more numerous and several fold more intense than the fewer weak sheath bands. Additional independent data demonstrating the axonal localization of the mRNA were obtained with *in situ* hybridization assays. They were initially based on the use of the generic probe [³H]poly(U) which allowed the detection of poly(A)RNA (Perrone Capano et al., 1987). In later analyses we used several specific riboprobes prepared from a library of axoplasmic RNA sequences (Chun et al., 1995; Gioio et al., 1994; Kaplan et al., 1992).

The identification of axoplasmic mRNAs (Giuditta, 1983; 1986) allowed the preparation of the corresponding cDNAs which were used to further examine the sequence complexity of axonal mRNA using hybridization techniques (Perrone Capano et al., 1987). The latter assays confirmed the occurrence of an axoplasmic population of mRNAs and estimated their diversity to be in the range of approximately 200 different mRNA species. In comparison, the mRNA diversity of the cell bodies of the giant axon was estimated at

10,000 different species. The availability of axoplasmic cDNA allowed furthermore the preparation of a library from which clones were isolated and sequenced, and compared to those of the cell bodies of the giant axon. The analyzed axoplasmic mRNAs included those coding for β-actin and β-tubulin (Kaplan et al., 1992), the heavy chain of kinesin (Gioio et al., 1994), enolase (Chun et al., 1995), and a novel peptide homologous to a peptide of the calcium channel of the sarcoplasmic reticulum (unpublished data). Conversely, a search for an axoplasmic mRNA coding for the voltage-dependent sodium channel yielded negative results, while the mRNA was identified in the library prepared from the cell bodies of the giant axon (unpublished data). In general, the sequence homology of axoplasmic and peri-karyal mRNAs was very high. An interesting difference emerged from the analysis of the 5′-untranslated region of β-tubulin which contained a 54 nt region present only in the ax-onal mRNA (Kaplan et al., 1992). This observation was tentatively interpreted to indicate the presence of a signal required for the axonal delivery of β-tubulin mRNA. However, a similar sequence was not detected in other axoplasmic mRNAs.

The axoplasmic localization of all the soluble components required for the activity of eukaryotic polysomes and of the main RNAs involved in protein synthesis (i.e. tRNAs, rRNAs and mRNAs), strongly suggested that active eukaryotic polysomes were present in the axoplasm of the giant axon. It could still be thought, however, that all these RNA and protein components were not involved in the process of protein synthesis, but were some-how kept inactive by the lack of some key factor or by the presence of an inhibitor. Al-though unlikely, this possibility was examined in an experiment which aimed at the isolation of active axoplasmic polysomes, i.e. of polysomes bearing nascent peptide chains.

To minimize the damage to the giant axon necessarily associated with the dissection procedure, the corresponding stellate nerve was separated from the squid mantle while preserving its continuity with the stellate ganglion. In addition, at its main bifurcation point, where further dissection is next to impossible, the most distal region of the mantle was left attached to the stellate nerve, albeit separate from the chamber holding the stellate nerve and ganglion. After incubation of the latter two compartments with [^{35}S]methionine for 1h, the axoplasm was extruded, homogenized in a suitable medium and treated as a post-nuclear fraction. Putative polysomes were purified by sedimentation of the axoplasm homogenate through a layer of 2M sucrose and then displayed by centrifugation on a 15–45% linear sucrose gradient.

Following the above procedure, radioactive proteins were detected in the polysomal region of the gradient. This pattern was markedly modified by treatment of the purified polysomes by RNAse or EDTA. Under these conditions most labelled peptides were re-covered from the lighter regions of the gradient (Giuditta et al., 1991), according to the well known disaggregating effects of RNAse and EDTA on the structure of polysomes and ribosomes, respectively. On the whole, the data demonstrated that the radiolabelled proteins recovered from the polysomal gradient were nascent peptides associated with ac-tive polysomes. Comparable results were obtained with polysomes prepared from control samples (giant fibre lobe and axonal sheath).

The isolation of axonal polysomes using a biochemical procedure was followed by the demonstration of axonal polysomes by morphological analyses of the giant axon with the method of electron spectroscopic imaging (ESI). This technique allows the selective detection of phosphorus atoms on the basis of the amount of energy loss encountered by impinging electrons. As ribosomes contain more than two thousands phosphorus atoms, they are readily identified as bright spots of the correct size over a dark background. Us-ing this technique large clusters of polysomes were detected in post-synaptic axonal sites (Martin et al., 1989) and in cortical layers of distal segments of the giant axon (Giuditta et al., 1991). The above data provided the first demonstration of the presence of eukaryotic polysomes engaged in protein synthesis in a mature axon. As a result, the synthesis of axonal proteins may not be exclusively attributed to the nerve cell bodies or to periaxonal glial cells, as axonal proteins are also formed by axonal polysomes.

A number of relevant questions were raised by this conclusion. They concerned the identity of the proteins synthesized by axoplasmic polysomes, the cellular origin of the components of the axonal system of protein synthesis (notably, of the tRNAs, rRNAs and mRNAs), and the generability of these findings to axons and nerve endings of other animal species.

2.2. Identification of the Proteins Synthesized by Axonal Polysomes

To identify the proteins synthesized by axonal polysomes, we followed two independent procedures, respectively based on the immunoabsorption of newly-synthesized proteins, or on the identification of the mRNAs associated with axoplasmic polysomes. The former method was applied to the study of axoplasmic samples obtained from giant axons incubated with [^{35}S]methionine, as well as to investigations of whole giant axons or stellate nerves. A meaningful analysis of the latter two samples could only concern neuron-specific proteins, as proteins with a wider cellular distribution could obviously be derived from periaxonal glial cells. Accordingly, our initial approach concerned the synthesis of neurofilament proteins (NF) by the isolated giant axon or the stellate nerve. Using squid specific NF antibodies, our analyses indicated that several NF proteins are synthesized by the giant axon or stellate nerve, including the 200 kDa, 70 kDa and 60 kDa NF proteins. A similar but not identical pattern was obtained from the giant fibre lobe (unpublished data).

Our alternative approach was based on the identification of the axoplasmic mRNAs associated with axoplasmic polysomes. The latter analyses were carried out with the technique of reverse transcriptase polymerase chain reaction (RT-PCR). They yielded positive results with regard to the axoplasmic mRNAs coding for β-actin and β-tubulin (Kaplan et al., 1992) and for kinesin (Gioio et al., 1994).

The relative abundance of several mRNAs in the giant axon and in its parent cell bodies has recently been examined using a quantitative RT-PCR method. These analyses confirmed that the levels of single axoplasmic mRNAs are several-fold lower than in the corresponding nerve cell bodies (Chun et al., in preparation). In addition, they indicated that the percent ratio of the axon versus the cell bodies markedly differs in different mRNAs, ranging from 12% with regard to kinesin mRNA to less than 1% with regard to enolase mRNA. As previously mentioned, an extreme example of this difference is provided by the mRNA coding for the voltage-dependent sodium channel, present in the cDNA library of the cell bodies but lacking in the axoplasmic library (unpublished data). These data demonstrate that the population of axonal mRNAs is qualitatively different from the population of cell bodies mRNAs and, as such, it cannot be considered derived from a random transport of perikaryal mRNAs into the axon. This conclusion is in agreement with the presence of a much higher number of different mRNAs in the cell bodies than in the axoplasm (10,000 versus a few hundreds; Perrone Capano et al., 1987).

2.3. Origin of Axonal RNA

In principle, the axonal RNAs may be thought to derive from transcription events occurring in the neuronal cell bodies or in the periaxonal glial cells, but not from axonal mitochondria, since they are cytoplasmic RNAs which may only be transcribed from nuclear DNA. In either case, the transcripts need to be transported to the axon, either by axoplasmic flow (if they derive from neuronal perikarya), or by intercellular transport (if they derive from periaxonal glial cells). We have approached this problem in the squid giant axon and have examined both possible sources of axonal RNA. While the evidence for a perikaryal origin of axonal RNA remains somewhat ambiguous, results are more convincing with regard to a local process of RNA synthesis (Cutillo et al., 1983).

Most experiments concerned the isolated giant axon incubated *in vitro* with [³H]uridine. Under these conditions radioactive RNA appeared in the extruded axoplasm in amounts which were much more than twice after 8h than after 4h (Cutillo et al., 1983). This observation suggested the existence of a lag period presumably required for the transport of newly-synthesized RNA into the axoplasmic core. Radioactive axoplasmic RNA was later shown to contain tRNA and rRNA (identified by electrophoretic analyses), and a sizable fraction of poly(A)RNA, presumably mRNA, identified by chromatography on oligo(dT)cellulose (Rapallino et al., 1988). The latter results confirmed that newly-synthesized axoplasmic RNA was cytoplasmic RNA (rather than mitochondrial RNA) and, therefore, that it required nuclear DNA for transcription. In turn, this suggested that it was derived from the nuclei of periaxonal glial cells which are the only cells surrounding the axon.

Labelled RNA was also detected in the axoplasm of the giant axon one day after injection of [³H]uridine in the most medial stellate nerve of a live squid, at a distance of about 5 cm from the stellate ganglion (Cutillo et al., 1983). While the peak of radioactive RNA coincided with the site of precursor injection, essentially no radioactive RNA was present in axoplasmic samples located a few centimeters away from the injection site in either direction, nor in the giant fibre lobe. These observations indicated that, in the live squid, axoplasmic RNA is synthesized locally, presumably in periaxonal glial cells.

More recent experiments carried out with isolated stellate nerves incubated with [³H]uridine have shown that the electrophysiological stimulation of the giant axon induces a significant increment in the amount of radioactive RNA recovered in the axoplasm (unpublished data). Comparable results have been obtained from perfused giant axons. In the latter preparation, the rate of appearance of radioactive RNA in the perfusate was found to be markedly increased by electrical stimulation (unpublished data). These results suggest that the stimulated axon releases one or more transmitters which bind to glial receptors, and induce an increase in the rate of glial RNA synthesis and/or transport into the axon. This interpretation is in agreement with a consistent body of data indicating the presence of several types of receptors in the periaxonal glial cells of the giant axon (Evans et al., 1986; Lieberman, 1991).

3. PROTEIN SYNTHESIS IN THE LARGE PRESYNAPTIC TERMINALS OF THE SQUID PHOTORECEPTOR NEURON

The second part of this article will be devoted to the presentation of experimental data indicating the presence of an eukaryotic system of protein synthesis in the large presynaptic endings of the squid photoreceptor neuron. The possibility of local protein synthesis in presynaptic terminals was actively considered in the late sixties and early seventies (Austin and Morgan, 1967; Morgan and Austin, 1968; Autilio et al., 1968; Cotman and Taylor, 1971; Gambetti et al., 1972; Gilbert, 1972; Ramirez et al., 1972) in experiments based on the incorporation of radiolabelled aminoacids in the proteins of synaptosomal fractions from mammalian brain. The initial encouraging results were progressively weakened by the consideration that synaptosomal fractions may contain a majority of cellular fragments other than presynaptic endings. As a result, their protein synthetic activity could hardly be attributed to the nerve terminals (Wedege et al., 1977; Verity et al., 1981; Irwin, 1985; Cheung, 1989; Rao and Steward, 1991). An additional unfavorable context was provided by the prevailing opinion that the axon lacks a local system of protein synthesis.

Prompted by the positive results obtained with the squid giant axon (Giuditta et al., 1983; 1986; 1991), we approched this general issue by using as a model system the synaptosomal fraction from squid optic lobes. This choice was suggested by the unusually high rate of protein synthesis of the fraction (Hernandez et al., 1976), and by its preparation as a floating layer (rather than as a sediment) which insured a better preservation of the synaptosomal particulates. An additional advantage consisted in the lack of dendrites in invertebrate neurons, as synapses are all of the axo-axonic type. As a result, most post-synaptic elements appear devoid of cytoplasm, presumably lost from unsealed axonal segments during the homogenization step (Figure 2).

As expected from brain particulates whose protein synthetic activity is assayed by the incorporation of a labelled aminoacid (i.e., without the addition of soluble factors), the reaction catalyzed by the squid synaptosomal fraction was fully blocked by a hypo-osmotic treatment and fully insensitive to a treatment with RNAse (Hernandez et al., 1976; Crispino et al., 1993a). These features confirmed that the active synaptosomal polysomes were surrounded by an intact plasma membrane which also engulfed all the soluble factors required for protein synthesis. The reaction was furthermore strongly inhibited by cycloheximide, a specific inhibitor of the cytoplasmic system of protein synthesis, and only partially inhibited by chloramphenicol, a specific inhibitor of mitochondrial protein synthesis (Hernandez et al., 1976; Crispino et al., 1993a).

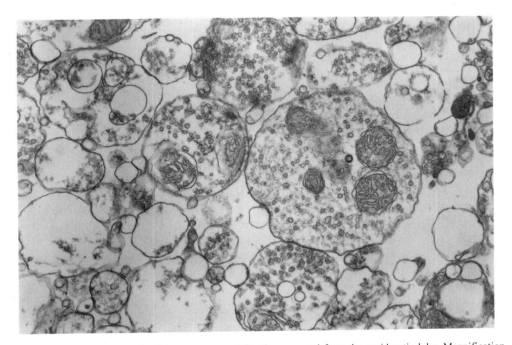

Figure 2. Electron micrograph of the synaptosomal fraction prepared from the squid optic lobe. Magnification, 14,000x.

To ascertain the degree of contamination of the synaptosomal fraction by cellular fragments derived from the nerve cell bodies or from glial cells, the electrophoretic pattern of the proteins synthesized by the synaptosomal fraction was compared with those of model systems of nerve cells (the giant fibre lobe) and glial cells (the stellate nerve). As these three patterns were markedly different from each other (Crispino et al., 1993a), the data suggested that the synaptosomal protein synthetic activity was not due to contaminating fragments derived from these cell types, but reflected the presence of a separate particulate system. This conclusion was strengthened by immunoabsorption analyses which demonstrated that the translation products of the synaptosomal fraction were markedly different from those of the giant fibre lobe. Notably, the newly-synthesized 70 kDa NF protein was prevalent over the 60 kDa NF protein in the synaptosomal fraction while the reverse was true in the giant fibre lobe (Crispino et al., 1993b). Glial cells were not utilized in these analyses on the ground that NF proteins are neuron-specific. The possibility that the above differences were due to a peculiar translation profile of the giant fibre lobe, possibly differing from that of the neuronal perikarya of the optic lobe, was considered unlikely, but was nonetheless addressed in comparable immunoabsorption analyses of the translation products of synaptosomal and microsomal polysomes of the optic lobe. The microsomal polysomes were considered to be derived from the free and membrane-bound polysomes of neuronal perikarya. The latter analyses confirmed our previous results in that they indicated that the 70 kDa NF protein was again prevalent over the 60 kDa NF protein in the translation products of synaptosomal polysomes but not of microsomal polysomes (Crispino et al., in preparation).

The above data supported the view that synaptosomal protein synthesis was not due to fragments of glial or nerve cell bodies, but reflected the activity of neuronal processes sheared during the homogenization step. In view of the lack of dendrites in squid neurons, and of the prevalence of large presynaptic terminals in the synaptosomal fraction (Figure 2), it seemed likely that the latter particulates were responsible for synaptosomal protein synthesis (Crispino et al., 1993a; 1994).

This view was confirmed by ESI inspection of the synaptosomal fraction which showed the presence of intraterminal phosphorus signals corresponding in size and spectroscopic properties to the signals emitted by ribosomes and polysomes of the neuronal cell body (Crispino et al., in preparation). As the large terminals of the optic lobe are derived from the retinal photoreceptors, comparable analyses were later extended to this neuronal type, by separately examining its cell body, the axon (optic nerve), and the large terminal "bag" located in the cortical layer of the optic lobe. These analyses confirmed that ESI signals corresponding in size and spectroscopic properties to those emitted by *bona fide* ribosomes and polysomes of the nerve cell body were present in the axons of the optic nerve and in its large presynaptic "bags" (Martin et al., in preparation). Interestingly, comparable ESI signals were not detected in other regions of the optic lobe neuropil.

Further proof of the presence of intraterminal eukaryotic polysomes was recently obtained by autoradiographic analyses of synaptosomal fractions incubated 1h with [^{35}S]methionine (Crispino et al., in preparation). Observation of the labelled fraction at the light microscope level indicated the presence of a diffuse background of radioactive proteins and of several clusters of silver grains with a diameter of a few μm. Both signals were drastically reduced following incubation of the synaptosomal fraction with cycloheximide. The latter observation indicated that eukaryotic polysomes were responsible for the incorporation reaction. Examination of the synaptosomal fraction at the electron microscope level showed that the clusters of silver grains were present over large presynaptic terminals identified by their content of synaptic vescicles and mitochondria. On the whole, these results support the conclusion that the large nerve endings of the squid photoreceptor neuron contain a population of active eukaryotic polysomes.

4. CONCLUSION

The data presented in this paper demonstrate that, at variance with the prevailing opinion, a local system of active eukaryotic polysomes is present in the squid giant axon and in the axons and presynaptic terminals of the squid photoreceptors. As a result, in addition to a delivery of proteins synthesized in the nerve cell body, these neuronal compartments may rely on a local system of protein synthesis. A similar activity is present in the large Mauthner axon of the goldfish (Koenig and Martin, 1996), and is likely to be of much wider distribution (Koenig, 1984; Giuditta et al., 1990). It remains to be seen whether the mRNAs recently detected in several vertebrate and invertebrate axons (Litman et al., 1994; Wensley et al., 1995; Olink-Coux et al., 1996; for reviews see Wiehelm and Vale, 1993; van Minnen, 1994) are involved in protein synthesis.

The demonstration of active eukaryotic polysomes in axons and in presynaptic regions opens the door to a renewed interest in the cellular and molecular mechanisms regulating neuronal protein synthesis and gene expression in the more general context of the plasticity of neuronal circuits.

5. SUMMARY

At variance with a prevailing opinion, active eukaryotic polysomes have been demonstrated in the squid giant axon and in the axons and presynaptic terminals of the squid photoreceptors. In addition, the axoplasm of the squid giant axon has been shown to contain a family of mRNAs, sizable amounts of rRNA, and all tRNAs and soluble factors required for extramitochondrial protein synthesis. Hence, these neuronal compartments should be thought of as relying on a local system of protein synthesis, besides the delivery of proteins synthesized in the neuronal soma. The occurrence of a local process of protein synthesis has been recently confirmed in the Mauthner axon of the goldfish, and may be of much wider distribution. On the other hand, the role of the various axonal mRNAs detected in several other vertebrate and invertebrates systems still remains to be clarified.

The presence of active polysomes in some types of axons and presynaptic terminals should elicit a renewed interest in the cellular and molecular mechanisms regulating protein synthesis in extrasomatic neuronal compartments, notably with regard to the more general issue of the processes responsible for the plasticity of neuronal circuits.

6. REFERENCES

Austin, L., & Morgan, I.G. (1967). Incorporation of 14-C-labelled leucine into synaptosomes from rat cerebral cortex "in vitro". J. Neurochem. 14:377–387.

Autilio, L.A., Appel, S.H., Pettis, P., Gambetti. P.G. (1968). Biochemical studies of synapses "in vitro". I. Protein synthesis. Biochem. 7:2615–2622.

Cheung, M.K. (1989). The specificity of glutamate inhibition of protein synthesis in synaptosomal fractions from rat cerebral cortex. Neurochem. Int. 15: 293–300.

Chun, J.T., Gioio, A.E., Crispino, M., Giuditta, A., & Kaplan, B.B. (1995). Characterization of squid enolase mRNA: sequence analysis, tissue distribution, and axonal localization. Neurochem. Res. 20:923–930.

Cotman, C.W., & Taylor, D.A. (1971). Autoradiographic analysis of protein syntesis in synaptosomal fractions. Brain Res. 29:366–372.

Crispino, M., Castigli, E., Perrone Capano, C., Martin, R., Menichini, E., Kaplan, B.B., & Giuditta, A. (1993a). Protein synthesis in a synaptosomal fraction from squid brain. Mol. Cell. Neurosci. 4:366–374.

Crispino, M., Perrone Capano, C., Kaplan, B.B., & Giuditta, A. (1993b). Neurofilament proteins are synthesized in nerve endings from squid brain. J. Neurochem. 61:1144–1146.

Crispino, M., Perrone Capano, C., Kaplan, B.B., & Giuditta, A. (1994). Squid optic lobe synaptosomes: what can they tell us about presynaptic protein synthesis? J. Neurochem. 63:387–389.

Cutillo, V., Montagnese, P., Gremo, F., Casola, L., & Giuditta, A. (1983). Origin of axoplasmic RNA in the squid giant fibre. Neurochem. Res. 8:1621–1634.

Evans, P.D., Reale, V., & Villegas, J. (1986). Peptidergic modulation of the membrane potential of the Schwann cell of the squid giant nerve fibre. J. Physiol. 379:61–82.

Gainer, H., Tasaki, I., & Lasek, R.J. (1977). Evidence for the glia neuron transfer hypothesis from intracellular perfusion studies of squid giant axons. J. Cell Biol. 74:524–530.

Gambetti, P., Autilio-Gambetti, L.A., Gonatas, N.K., & Shafer, B. (1972). Protein synthesis in synaptosomal fractions. J. Cell Biol. 52:526–535.

Gilbert, M.J. (1972). Evidence for protein synthesis in synaptosomal membranes. J. Biol. Chem. 247:6541–6550.

Gioio, A.E., Chun, J.T., Crispino, M., Perrone Capano, C., Giuditta, A., & Kaplan, B.B. (1994). Kinesin mRNA is present in the squid giant axon. J. Neurochem. 63:13–18.

Giuditta, A. (1980). Origin of axoplasmic protein in the squid giant axon. Riv. Biol. 73:35–49.

Giuditta, A., Dettbarn, W.D., & Brzin, M. (1968). Protein synthesis in the isolated giant axon of the squid. Proc. Natl. Acad. Sci. USA 59:1284–1287.

Giuditta, A., Metafora, S., Felsani, A., & Del Rio, A. (1977). Factors for protein synthesis in the axoplasm of the squid giant axons. J. Neurochem. 28:1393–1395.

Giuditta, A., Cupello, A., & Lazzarini, G. (1980). Ribosomal RNA in the axoplasm of the squid giant axon. J. Neurochem. 34:1757–1760.

Giuditta, A., Hunt, T., & Santella, L. (1983). Messenger RNA in squid axoplasm. Biol. Bull. 165:526.

Giuditta, A., Hunt, T., & Santella, L. (1986). Messenger RNA in squid axoplasm. Neurochem. Intern. 8:435–442.

Giuditta, A., Menichini, E., Castigli, E., & Perrone Capano, C. (1990). Protein synthesis in the axonal territory. In Giuffrida Stella, A.M., de Vellis, J., Perez Polo, R. (eds.): "Regulation of Gene Expression in the Nervous System." New York: Alan Liss, Inc., pp 205–218.

Giuditta, A., Menichini, A., Perrone Capano, C., Langella, M., Martin, R., Castigli, E., & Kaplan, B.B. (1991). Active polysomes in the axoplasm of the squid giant axon. J. Neurosci. Res. 28:18–28.

Hernandez, A.G., Langford, G.M., Martinez, J.L., & Dowdall, J. (1976). Protein synthesis by synaptosomes from the head ganglion of the squid "Loligo Pealii". Acta Cient. Venez. 27:120–123.

Ingoglia, N.A., Giuditta, A., Zanakis, M.F., Babigian, A., Tasaki, I., Chakraborty, G., & Sturman, J. (1983). Incorporation of [^{3}H]aminoacids into proteins in a partially purified fraction of axoplasm: Evidence for transfer RNA mediated, post-translational protein modification in squid giant axon. J. Neurosci. 3:2463–2473.

Irwin, C.C. (1985). Comparison of protein syntesis in mitochondria, synaptosomes and intact brain cells. J. Neurochem. 44:433–438.

Kaplan, B.B., Gioio, A.E., Perrone Capano, C., Crispino, M., & Giuditta, A. (1992). β-actin and β-tubulin are components of a heterogeneous mRNA population present in the squid giant axon. Mol. Cell. Neurosci. 3:133–144.

Koenig, E. (1979). Ribosomal RNA in Mauthner axon: implications for a protein synthesizing machinery in the myelinated axon. Brain Res. 174:95–107.

Koenig, E. (1984). Local synthesis of axonal proteins, in "Handbook of Neurochemistry", Vol. 7 (Lajtha, A. ed.), 2nd edition, pp. 315–340. Plenum Press, New York.

Koenig, E. (1991). Evaluation of local synthesis of axonal proteins in the goldfish Mauthner cell axon and axons of dorsal and ventral roots of the rat *in vitro*. Mol. Cell. Neurosci. 2:384–394.

Koenig, E., & Koelle, G.B. (1960). Acetylcholinesterase regeneration in peripheral nerve after irreversible inactivation. Science 132:1249–1250.

Koenig, E., & Martin, R. (1996). Cortical plaque-like structures identify ribosome-containing domains in the Mauthner cell axon. J. Neurosci. 16:1400–1411.

Lasek, R.J., Dabrowski, C., & Nordlander, R. (1973). Analysis of axoplasmic RNA from invertebrate giant axons. Nature NB 244:162–165.

Lasek, R.J., Gainer, H., & Przybylski, K. (1974). Transfer of newly-synthesized proteins from Schwann cells to the squid giant axon. Proc. Natl. Acad. Sci. USA 71:1188–1192.

Lasek, R.J., Gainer, H., & Barker, J.L. (1977). Cell to cell transfer of glial proteins to the squid giant axon. J. Cell Biol. 74:501–523.

Lieberman, E.M. (1991). Role of glutamate in axon-Schwann cell signaling in the squid. Ann. New York Acad. Sci. 633:448–457.

Litman, P., Barg, J., & Ginzburg, I. (1994). Microtubules are involved in the localization of tau mRNA in primary neuronal cells in culture. Neuron 13:1463–1474.

Martin, R., Fritz, W., & Giuditta, A. (1989). Visualization of polyribosomes in the postsynaptic area of the squid giant synapse by electron spectroscopic imaging. J. Neurocytol. 18:11–18.

Morgan, I.G., & Austin, L. (1968). Synaptosomal protein synthesis in a cell-free system. J. Neurochem. 15:41–51.

Olink-Coux, M., & Hollenbeck, P.J. (1996). Localization and active transport of mRNA in axons of sympathetic neurons in culture. J. Neurosci. 16:1346–1358.

Perrone Capano, C., Gioio, A.E., Giuditta, A., & Kaplan, B. (1987). Occurrence and sequence complexity of polyadenylated RNA in squid axoplasm. J. Neurochem. 49:698–704.

Ramirez, G., Levitan, I.B., & Mushynski, E. (1972). Highly purified synaptosomal membranes from rat brain: incorporation of amino acids into membrane proteins *in vitro*. J. Biol. Chem. 247:5382–5390.

Rao, A., & Steward, O. (1991). Evidence that protein costituents of postsynaptic membrane specializations are locally synthesized: analysis of proteins synthesized within synaptosomes. J. Neurosci. 11:2881–2895.

Rapallino, M.V., Cupello, A., & Giuditta, A. (1988). Axoplasmic RNA species synthesized in the isolated squid giant axon. Neurochem. Res. 13: 625–631.

van Minnen, I. (1994). RNA in axonal domain: a new dimension in neuronal functioning? Histochem. J. 26:377–391.

Verity, M.A., Cheung, M.K., & Brown, W.J. (1981). Studies on valinomycin inhibition of synaptosome-fraction protein synthesis. Biochem. J. 196:25–32.

Wedege, E., Luqmani, Y., & Bradford, H.F. (1977). Stimulated incorporation of amino acids into proteins of synaptosomal fractions induced by depolarizing treatments. J. Neurochem. 29:527–537.

Wensley, C.H., Stone, D.M., Baker, H., Kauer, J.S., Margolis, F.L., & Chikaraishi, D.M. (1995). Olfactory marker protein mRNA is found in axons of olfactory receptor neurons. J. Neurosci. 15:4827–4837.

Wiehelm, J.E., & Vale, R.D. (1993). RNA on the move: the mRNA localization pathway. J. Cell Biol. 123:269–274.

13

PROTEIN SYNTHESIS IN PRESYNAPTIC ENDINGS OF SQUID BRAIN: REGULATION BY Ca^{2+} IONS

Juan Claudio Benech,[1,2] Mariana Crispino,[4] Barry. B. Kaplan,[3] and Antonio Giuditta[4]

[1]División Biofísica
Instituto de Investigaciones Biológicas Clemente Estable
Avenida Italia 3318, CP 11600
Montevideo, Uruguay
[2]Area Biofísica, Facultad de Veterinaria
Av. Lasplaces 1550
Montevideo, Uruguay
[3]Western Psychiatric Institute and Clinic
University of Pittsburgh Medical Center
3811 O'Hara Street
Pittsburgh, Pennsylvania 15213-2593, USA
[4]Department of General and Environmental Physiology
University of Naples "Federico II"
Via Mezzocannone 8
80123, Naples, Italy

1. INTRODUCTION

It is generally accepted that long-term synaptic plasticity induced by electrophysiological or behavioral stimulation requires the modulation of gene expression, eventually leading to modification of the set of synaptic proteins (Montarolo et al., 1986; Otani et al., 1989). These changes are believed to occur in presynaptic and postsynaptic sites (dendrites) of the neuron. The ability of dendrites to synthesize proteins, is largely accepted (Rao & Steward, 1991). On the other hand, the concept of protein synthesis in presynaptic terminals is still controversial. This possibility was considered in the past (Austin et al., 1967; Morgan & Austin, 1968; Gordon & Deanin, 1968; Bridgers et al., 1971; Cotman & Taylor, 1971), but failed to achieve acceptance as the protein synthetic activity of synaptosomal fractions could be attributed to contaminating fragments of glial cells or dendrites, rather than to nerve endings.

During the last 15 years, the existence of a local system of protein synthesis in the axonal domain of the neuron (proposed by Koenig & Koelle, 1960) has found increasing

Calcium and Cellular Metabolism: Transport and Regulation, edited by Sotelo and Benech.
Plenum Press, New York, 1997

155

support in literature data mainly concerning the Mauthner axon of the goldfish (Koenig, 1979; 1991; Koenig & Martin, 1996) and the giant axon of the squid (see chapter 12), but also dealing with mammalian axons (Koenig, 1967a, 1967b, 1967c; Benech et al., 1968; Tobias & Koenig; 1975a, 1975b; Frankel & Koenig, 1977; 1978; Benech, et al. 1982; Alvarez & Benech, 1983; Sotelo et al., 1992; for reviews see Koenig (1984) and Alvarez & Torres (1985)).

Recently, an extramithocondrial system of protein synthesis was described in the synaptosomal fraction of the squid optic lobe (Crispino et al., 1993a; Crispino et al., 1993b). Several lines of evidence (Crispino et al., 1997; Martin et al., 1997) indicated that this activity is mainly present in the large nerve endings which are the prevalent components of the synaptosomal fraction (see chapter 12). As the large nerve terminals of the optic lobe only derive from retinal photoreceptors (Young, 1962; Cohen, 1973; Young, 1974; Haghighat, 1984), the latter neurons have confidently been identified as the source of the active nerve terminals recovered in the synaptosomal fraction. The presence of active eukaryotic polysomes in model axons and presynaptic terminals raises several questions with regard to their role and mechanisms of regulation. In the present report, we will present evidence that in synaptosomes from the squid optic lobe, presynaptic protein synthesis is strongly modulated by the intraterminal concentration of calcium ions (Benech, J. C. et al., 1994). In addition, we will describe data suggesting that presynaptic protein synthesis is increased following *in vivo* stimulation of the retinal photoreceptors by light (Benech, J. C. et al., 1996).

2. EFFECT OF THE IONIC COMPOSITION OF THE INCUBATION MEDIUM

The protein synthetic activity of the squid synaptosomal fraction is strongly dependent on the ionic composition of the incubation medium (Crispino et al., 1993a). In fact, the rate of the reaction decreases by approximately two thirds when the standard medium (artificial sea water, ASW) is substituted by an iso-osmolar sucrose solution and ions are omitted. Maximal activity may again be regained by the addition of physiological concentration of Na^+ and K^+ to the incubation medium.

The first indication of an inhibitory effect of calcium ions was obtained when the rate of $[^{35}S]$methionine incorporation into proteins was measured in Mg^{2+} free ASW. Under this condition, raising the Ca^{2+} concentration from 11 mM (its normal value in ASW) to 16 mM or more induced an inhibition of 24% ($p<0.005$; $n=11$; Student t-test).

Conversely, when the rate of the reaction was measured in Ca^{2+} free ASW, a significant activation was observed (28%; $p<0.003$; $n=6$; Student t-test). The latter effect depended of the presence of Mg^{2+}, as the simultaneous omission of Ca^{2+} and Mg^{2+} did not yield an activation.

3. MANIPULATION OF THE INTRASYNAPTOSOMAL $[Ca^{2+}]$

Relatively simple procedures may be used to modify the intrasynaptosomal Ca^{2+} concentration. In the control condition, the cytosolic concentration of Ca^{2+} is presumably low and the concentration of Ca^{2+} sequestered in the internal stores is high. When synaptosomes are incubated in a reaction medium lacking Ca^{2+} (in presence of 1mM EGTA), with or without Ca^{2+} ionophores, a "Ca^{2+} depleted state" is attained (Brostrom & Brostrom, 1990) characterized by low cytosolic and low sequestered Ca^{2+}levels. In a third condition, i.e. when synaptosomes are incubated in ASW containing a Ca^{2+} ionophore, a high Ca^{2+} concentration is induced in the cytosol and in the internal stores, as they approach the concentration of Ca^{2+} in the incubation medium.

3.1. Raising the Concentration of Cytosolic Ca^{2+}

The rate of synaptosomal protein synthesis measured in ASW containing the calcium ionophore A23187 was strongly inhibited (Crispino et al., 1993). This initial observation, was confirmed using ionomycin, a chemically different calcium ionophore (Liu & Herman, 1978; Toeplitz et al., 1979). The degree of inhibition was dependent on the concentration of either ionophore (data not shown). Maximal inhibition achieved at 1 μM in either case (Table 1). The inhibition induced by A23187 (70% p<0.01; n=5 Student t-test) was higher than the inhibition observed with ionomycin (54%; p<0.001; n=6; Student t-test).

As shown in Table 1, when the rate of synaptosomal protein synthesis was measured in Ca^{2+}-free ASW (irrespective of the presence of 1mM EGTA), containig 1 μM of A23187, the degree of inhibition remained high (69%; p<0.002; n=5; Student t-test). A somewhat lower inhibition was induced by 1 μM ionomycin in Ca^{2+}-free ASW (38%; p<0.04; n=3; Student t-test). On the other hand, a stronger inhibition (higher than 70%) was observed when either ionophore (1 μM) was used in Mg^{2+}-free ASW. Interestingly, when protein synthesis was measured in Ca^{2+}-free, Mg^{2+}-free ASW, the inhibition induced by 1 μM of A23187 was markedly lower (30%). The degree of inhibition remained low even when the Ca^{2+} concentration was raised up to 1 μM, but strong inhibitory effects appeared at higher Ca^{2+} levels.

The stronger inhibition by A23187 observed in Ca^{2+}-free ASW as compared with Ca^{2+}-free, Mg^{2+}-free ASW suggested that a significant inhibitory effect was also exerted by Mg^{2+} (55 mM in ASW). In addition, the weaker but sizable inhibition observed in Ca^{2+}-free, Mg^{2+}-free ASW appeared to be due to the release of Ca^{2+} from internal stores as a result of ionophore binding to intrasynaptosomal membranes.

To explore the possibility of an involvement of internal Ca^{2+} stores in the regulation of synaptosomal protein synthesis, we tested the effect of thapsigargin (Sagara et al., 1991; Lytton et al., 1991), a highly specific inhibitor of the SERCA family of Ca^{2+} pumps (SarcoEndoplasmic Reticulum Ca^{2+} pumps). Incubation of synaptosomes with thapsigargin promoted a marked inhibition of protein synthesis that was almost maximal at a concentration of 10 μM, and was essentially the same in presence of the normal Ca^{2+} concentration (11 mM) or in its absence (1 mM EGTA added, Table 1).

Further support to the above possibility was provided by the marked inhibitory effect induced by caffeine, known to increase the outflow of Ca^{2+} from the sarcoplasmic reticulum into the cytosol as a result of binding to the ryanodin receptor (Endo et al.,

Table 1. Changes of synaptosomal protein synthesis activity related to the raising of cytosolic $[Ca^{2+}]$

Experimental condition	Percentage of control [³⁵S]-methionine incorporation	
	Normal ASW (11 mM Ca^{2+})	Ca^{2+}-free ASW (1 mM EGTA)
Control	100	100
A23187 (1 μM)	30	31
Ionomycin (1μM)	46	62
Caffeine (10 mM)	58	57
Thapsigargin (10 μM)	21	22

Table 2. Changes in synaptosomal protein synthesis activity
related to the lowering of cytosolic [Ca^{2+}], or hindering
the functional role of calcium

Experimental condition	Percentage of control [^{35}S]-methionine incorporation	
	Normal ASW (11 mM Ca^{2+})	Ca^{2+}-free ASW (1 mM EGTA)
Control	100	100
BAPTA-AM (20 µM)	20	—
W7 (100 µM)	17	10
TFP (50 µM)	14	10
Calphostin (20 µM)	15	—

1970; Inui et al., 1987; Erlich & Watras, 1988). The caffeine inhibition remained essentially the same in the absence of Ca^{2+} in the external medium (Table 1).

3.2. Lowering the Concentration of Cytosolic Ca^{2+} or Hindering Its Functional Role

To study the effect of lowering cytosolic Ca^{2+} concentration, synaptosomes were preloaded with BAPTA-AM for 15 minutes at 22°C, before incubation with [^{35}S]methionine for 1 hour. Under these conditions the rate of protein synthesis decreased progressively depending of the concentration BAPTA-AM (data non shown). Maximal inhibition (80%) occurred at 20µM BAPTA-AM (Table 2), and half maximal inhibition at 0.5 µM.

A similar effect was obtained using N-(6-aminoexil)-5chloro-1-naphthalene-sulfonamide (W7) and trifluoperazine (TFP). These two drugs inhibit calmodulin (Hidaka et al., 1981; Snelling & Nicholls, 1984; Schweitzer, 1987; Stiges & Talamo, 1993), a calcium binding protein that mediates the effect of calcium on protein kinase C and on calcium/calmodulin-dependent protein kinase II C (Wang et al., 1988). As shown in Table 2, synaptosomal protein synthesis measured in Ca^{2+}-free ASW (1mM EGTA) was inhibited by either W7 or TFP. The degree of inhibition was dose-dependent. Half maximal inhibition was observed at 30 µM W7 and 15 µM TFP (not shown). A Strong inhibition by either calmodulin antagonist was also observed in normal ASW.

As calmodulin regulates the activity of protein kinase C, we also tested the effect of calphostin, a specific inhibitor of this protein kinase (Kobayashi et al., 1989) which also exerted a marked inhibition on synaptosomal protein synthesis (Table 2).

4. EFFECT OF LIGHT STIMULATION ON LIVE SQUID

As mentioned above, the squid retinal photoreceptors have been identified as the main source of the large presynaptic terminals recovered in the synaptosomal fraction (Young, 1962; Cohen 1973, Young 1974; Haghighat 1984) which are responsible for a large proportion of its protein synthetic activity (Crispino et al. 1997; Martin et al. 1997). This unique situation prompted us to examine the effect of stimulation of the retinal photoreceptors on synaptosomal protein synthesis. To this aim, live squid were kept approximately seven hour in a tank with running sea water at a temperature of 18–20°C in a lighted environment, or conversely in the dark. The latter condition was obtained by covering one tank with several sheets of black plastic. Preparation of the synaptosomal fraction from "light-adapted" and "dark-adapted" squids and the determination of their protein synthetic activity revealed that the rate of the reaction was approximately 80% higher in synaptosomes prepaired from "light-adapted" animals. This effect was not likely to be due to circadian influences as the experiment was repeated with essentially similar results at different times of the day and of the night.

5. DISCUSSION

The data presented in this paper, support the notion that the protein synthetic activity of the large nerve endings of the squid photoreceptor neurons is modulated by Ca^{2+} and presumably by neuronal activity. The former conclusion is based on the results of experiments with different compounds known to interfere with calcium homeostasis, including two calcium ionophores and several others drugs. The data suggest that synaptosomal protein synthesis is close to its maximal rate at the normal concentration of cytosolic calcium ions, and is inhibited whenever this concentration is increased or decreased.

Strong evidence for the inhibitory effect of high concentrations of cytosolic calcium is provided by the observation that the calcium ionophores A23187 and ionomycin induced a marked dose-dependent inhibition of the synaptosomal protein synthesis which remains essentially the same when the reaction occurs in Mg^{2+}-free or Ca^{2+}-free ASW. The inhibitory effect observed in the latter condition is likely to reflect the release of Ca^{2+} from internal stores. A clear demonstration that the rate of synaptosomal protein synthesis depends on the concentration of Ca^{2+} is provided by the observation that in Mg^{2+}-free ASW containing 1 μM A23187, a progressive inhibition is induced by raising the level of calcium in the incubation medium. Under these conditions (i.e., in presence of Ca^{2+} ionophore), the Ca^{2+} concentration inside the synaptosomes approaches that in the medium. This experiment required the use of Mg^{2+}-free ASW to avoid the confounding inhibitory effect of this ion on synaptosomal protein synthesis. The role of internal synaptosomal stores in maintaining a low cytosolic calcium concentration (thereby modulating synaptosomal protein synthesis) is indicated by the strong inhibition induced by thapsigargin, a highly specific inhibitor of the SERCA family calcium pumps, responsible for the removal of calcium from the cytosol into the endoplasmic reticulum. In presence of thapsigargin, the cytosolic Ca^{2+} concentration is expected to increase. A similar effect was reported in HeLa cells (Preston & Berlin, 1992). The inhibition of synaptosomal protein synthesis promoted by caffeine in the absence of external Ca^{2+} is in agreement with this notion, as it strongly suggests that this effect is mediated by the release of calcium from internal stores through the calcium channel associated with the ryanodine receptor.

On the other hand, the conclusion that synaptosomal protein synthesis is likewise inhibited by decreasing the cytosolic Ca^{2+} concentration or by preventing the functional role of Ca^{2+} is based on the results of experiments using BAPTA-AM, a calcium-sequestering compound known to cross plasma membrane, and on data obtained with two calmodulin antagonists, TFP and W7. The latter data also suggest that presynaptic protein synthesis may be regulated by calmodulin.

That accumulation of BAPTA-AM within the cytosol results in a decrease of Ca^{2+} concentration has been demonstrated in different cell types (Muallem et al., 1990; Preston & Berlin, 1992). On the other hand, antagonists of calmodulin have been reported to inhibit protein synthesis in ascites tumor cells (Kumar et al., 1991). In brief, presynaptic protein synthesis is inhibited when the cytosolic calcium concentration goes below the level needed to activate calmodulin (as it happens in presence of BAPTA-AM) or following addition of compounds known to directly antagonize calmodulin, such as TFP and W7.

Results obtained in several cell types have suggested that sequestered calcium, rather than free cytosolic Ca^{2+}, is responsible for the maintainance of an optimal level of protein synthesis (Brostrom et al., 1983; Chin et al., 1988; Brostrom & Brostrom, 1990; Preston & Berlin, 1992). The present results do not directly support this view, but cannot exclude its validity.

It is important to emphazise that the modulatory role of Ca^{2+} described in this paper largely concerns the local system of protein synthesis present in the presynaptic endings of the squid retinal photoreceptors. In view of the well known participation of calcium ions in synaptic transmission, presynaptic changes in cytosolic Ca^{2+} concentration might have a regulatory role on presynaptic protein synthesis depending on the frequency of incoming stimuli and on the rate of removal of excess cytosolic Ca^{2+}. The difference in rate of protein synthesis observed in synaptosomes prepared from "light- or dark-adapted" squids might be attributed to differences in firing frequency of the retinal photoreceptors leading to changes in presynaptic Ca^{2+} concentration.

It is important to note that in the experiments in which calcium homeostasis was modified by drugs like the Ca^{2+} ionophores or thapsigargin, the increase in cytosolic Ca^{2+} was not transient, but persisted throughout the incubation period, at variance with the dynamic condition presumed to prevail *in vivo*. This difference might be relevant when the effect observed *in vivo* are compared to those observed *in vitro*. Determination of presynaptic Ca^{2+} levels using suitable indicators could be required to examine this possibility.

The presumed involvement of calmodulin and protein kinase C in the regulation of presynaptic protein synthesis raises the question of the identification of the protein substrate(s) whose phosphorilation may be required to attain an optimum rate of protein synthesis. Two candidates are the initiation factor eIF-2 (Kimball & Jefferson, 1991; Kimball & Jefferson, 1992) and the elongation factor eEF-2 (Nygard et al., 1991) whose activities decrease upon phosphorilation. Further work will be required to confirm this possibility.

6. SUMMARY

Large nerve terminals of retinal photoreceptor neurons were identified as the prevalent component of the synaptosomal fraction obtained from the optic lobe of the squid. Data reported in this paper, support the notion that the extramitochondrial protein synthesis activity of this synaptosomal fraction is modulated by Ca^{2+}, and presumably by neural activity.

The above proposed role of Ca^{2+} in modulating synaptosomal protein synthesis was tested using a set of different drugs (i.e. calcium ionophores, thapsigargin, BAPTA and the Calmodulin inhibitors W-7 and Trifluoperazine), already known to modify calcium homeostasis. The reported results suggest that synaptosomal protein synthesis is close to its optimal level at normal concentration of cytosolic calcium ions, and is inhibited whenever this concentration is increased or decreased.

To examine the effects of retinal photoreceptor neuron stimulation on synaptosomal protein synthesis, live squids were kept in light and dark conditions. Determination of protein synthesis levels in the synaptosomal fractions obtained from the "light" and "dark-adapted" animals indicated that protein specific activity was significantly higher (approximately 80%) in squids kept in a light environment.

7. REFERENCES

Alvarez, J., & Benech, C. R. (1983). Axoplasmic incorporation of aminoacids in myelinated fiber exceeds that of its soma: an autoradiographic study. Exp. Neurol. 82:25–42.

Alvarez, J., & Torres, J. C. (1985). Slow axoplasmic transport: a fiction? J. Theor. Biol. 112:627–651.

Austin, L., & Morgan, I.G. (1967). Incorporation of 14-C-labelled leucine into synaptosomes from rat cerebral cortex "in vitro". J. Neurochem. 14:377–387.

Benech, C.R., Saá, E.A., & Franchi, C.M. (1968). In vivo local uptake of C-14 orotic acid by peripheral nerve. Exp. Neurol. 22:436–443.

Benech, C. R., Sotelo, J.R., Menéndez, J., & Correa-Luna, R. (1982). Autoradiographic study of RNA and protein synthesis in sectioned peripheral nerves. Exp. Neurol. 76:72–82.

Benech, J.C., Crispino, M., Chun, J.T., Kaplan, B.B., & Giuditta, A. (1994). Protein Synthesis in nerve endings from squid brain: modulation by calcium ions. Biol. Bull. 187:269.

Benech, J.C., Crispino, M., Martin, R., Alvarez, J., Kaplan, B.B., & Giuditta, A. (1996). Protein synthesis in the presynaptic endings of the squid photoreceptor neuron: *in vitro* and *in vivo* modulation. Biol. Bull. 191:263.

Bridgers, W.F., Cunningham, R.D., & Gressett, G. (1971). Properties distinguishing mitochondrial and synaptosomal protein synthesis. Biochem. Biophys. Res. Commun. 45:351–357.

Brostrom, C.O., Bocckino, S.B., & Brostrom, M. (1983). Identification of a calcium requirement for protein synthesis in eukaryotic cells. J. Biol. Chem. 258:14390–14399.

Brostrom, C.O., & Brostrom, M. (1990). Calcium-dependent regulation of protein synthesis in intact mammalian cells. Annu. Rev. Physiol. 52:577–590.

Chin, K.V., Cade, C., Brostrom, M., Brostrom C.O. (1988). Regulation of protein synthesis in intact rat liver by calcium mobilizing agents. Int. J. Biochem. 20:1313–1319.

Cohen, A.I. (1973). An ultrastructural analysis of the photoreceptors of the squid and their synaptic conections. III. Photoreceptor terminations in the optic lobes. J. Comp. Neurol. 147:399–426.

Cotman, C.W., & Taylor, D.A. (1971). Autoradiographic analysis of protein synthesis in synaptosomal fractions. Brain. Res. 29:366–372.

Crispino, M., Castigli, E., Perrone Capano, C., Martin, R., Menichini, E., Kaplan, B.B., & Giuditta, A. (1993a). Protein synthesis in a synaptosomal fraction from squid brain. Mol. Cell. Neurosci. 4:366–374.

Crispino, M., Perrone Capano, C., Kaplan, B.B., & Giuditta, A. (1993b). Neurofilament proteins are synthesized in the nerve endings from squid brain. J. Neurochem. 61:1144–1146.

Crispino, M., Kaplan, B.B., Martin, R., Alvarez, J., Chun, J.T., Benech, J.C., & Giuditta, A. (1997). Active polysomes are present in the large presynaptic endings of the synaptosomal fraction from squid brain. Submitted.

Endo, M., Tanaka, M., & Ogawa, Y. (1970). Calcium induced release of calcium from the sarcoplasmic reticulum of skinned muscle fibres. Nature. 228:34–36.

Ehrlich, B.E., & Watras, J. (1988). Inositol 1,4,5-trisphosphate activates a channel from smooth muscle sarcoplasmic reticulum. Nature. 336:583–586.

Frankel, R.D., & Koenig, E. (1977). Identification of major indigenous protein components in mammalian axons and locally synthesized axonal proteins in hypoglossal nerve. Exp. Neurol. 57:282–295.

Frankel, R.D., & Koenig, E. (1978). Identification of locally synthesized proteins in proximal stumps axons of the neurotomized hypoglossal nerve. Brain. Res. 141:67–76.

Gordon, M.W., & Deanin, G.G. (1968). Protein synthesis by isolated rat brain mithocondria and synaptosomes. J. Biol. Chem. 243:4222–4226.

Haghighat, N., Cohen, R.S., & Pappas, G.D. (1984). Fine structure of squid (Loligo pealei) optic lobe synapses. Neurosci. 13:527–546.

Hidaka, H; Sasaki, Y; Tanaka, T; Endo, T; Ohno, S; Fujii, Y. & Nagata, T. (1981). N-(6-aminohexyl)-5-chloro-1-naphtalenesulfonamide, a calmodulin antagonist, inhibits cell proliferation. Proc Natl. Acad. Sci. USA. 78:4354

Inui, M., Saito, A., & Fleischer, S. (1987). Purification of the ryanodine receptor and identity with feet structures of junctional terminal cisternae of sarcoplasmic reticulum from skeletal muscle. J. Biol. Chem. 262:1740–1747.

Kimball, S. R., & Jefferson, L. (1991). Mechanism of inhibition of peptide chain initiation by amino acid deprivation in perfused rat liver. Regulation involving inhibition of eukaryotic initiation factor 2 alpha phosphatase activity. J. Biol. Chem. 266:1969–1974.

Kimball, S. R., & Jefferson, L. (1992). Regulation of protein synthesis by modulation of intracellular calcium in rat liver. Am. J. Physiol. 263 (Endocrinol. Metab. 26):E958-E964.

Kobayashi, E. (1989). Calphostin C (UCN-1028C), a novel microbial compound, is a highly specific inhibitor of protein kinase C. Biochem. Biophys. Res. Commun. 159:548–553.

Koenig, E., & Koelle, G.B. (1960). Acetylcholinesterase regeneration in peripheral nerve after irreversible inactivation. Science. 132:1249–1250.

Koenig, E. (1967a). Synthetic mechanisms in the axon. I. Local axonal synthesis of acetylcholinesterase. J. Neurochem. 12:343–355.

Koenig, E. (1967b). Synthetic mechanisms in the axon. III. Stimulation of acetylcholinesterase synthesis by actinomycin D in the hypoglossal nerve. J. Neurochem. 14:429–435.

Koenig, E. (1967c). Synthetic mechanisms in the axon. IV. In vitro incorporation of [3H]precursors into axonal proteins and RNA. J. Neurochem. 14:437–446.

Koenig, E. (1979). Ribosomal RNA in the Mauthner axon: implications for a protein synthesis machinery in the myelinated axons. Brain. Res. 174:95–107.

Koenig, E. (1984). Local synthesis of axonal proteins. In Lajtha, A. In: Handbook of Neurochemistry, 7:315–340, second edition. Plenum Press, New York.

Koenig, E. (1991). Evaluation of local synthesis of axonal proteins in the goldfish Mauthner cell axon and axons of the dorsal and ventral roots of the rat in vitro. Mol. Cell. Neurosci. 2:384–394.

Koenig, E., & Martin, R. (1996). Cortical plaque-like structure identify ribosome-containing domains in the Mauthner cell axon. J. Neurosci. 16:1400–1411.

Kumar, R. V; Panniers, R; Wolfman, A. & Henshaw, E. C. (1991). Inhibition of protein synthesis by antagonists of calmodulin in Ehrlich ascites tumor cells. Eur. J. Biochem. 195:313–319.

Liu, C., & Herman, T. E. (1978). Characterization of ionomycin as a calcium ionophore. J. Biol. Chem. 253:5892–5894.

Lytton, J., Westlin, M. & Hanley, M.R. (1991). Thapsigargin inhibits the sarcoplasmic or endoplasmic reticulum Ca2+ATPase family of calcium pumps. J. Biol. Chem. 266:17067–17071.

Martin, R., Crispino, M., Kaplan, B.B., & Giuditta, A. (1997). Ribosome-like particles are present in axons and presynaptic endings of the squid photorreceptor neuron. Submitted.

Morgan, I.G., & Austin, L. (1968). Synaptosomal protein synthesis in a cell-free system. J. Neurochem. 15:41–51.

Montarolo, P.G., Golet, P., Castelluci, V.F., Morgan, J., Kandel, E.R., & Schacher, S. (1986). A critical period for macromolecular synthesis in long-term heterosynaptic facilitation in Aplisia. Science 234:1249–1254.

Muallem, S. Khademazad, M. & Sachs, G. (1990). The route of Ca^{2+} entry during reloading of the intracellular Ca^{2+} pool in pancreatic acini. J. Biol. Chem. 265:2011–2016.

Nygard, O.; Nilsson, A.; Carlberg, U.; Nilson, L. & Amons, R. (1991). Phosphorylation regulates the activity of the eEF-2 specific Ca^{2+} - and calmoldulin-dependent protein kinase III. J. Biol. Chem. 266:16425–16430.

Otani, S., Marshall, C.J., Tate, W.P., Goddard, G.V., & Abraham, W.C. (1989). Maintenance of long-term potentiation in rat dentate gyrus requires protein synthesis but not messenger RNA synthesis immediately posttetanization. Neuroscience 28:519–526.

Preston, G.F., & Berlin, R.D. (1992). An intracellular calcium store regulates protein synthesis in HeLa cells, but is not the hormone-sensitive store. Cell Calcium. 13:303–312.

Rao, A., & Steward, O. (1991). Evidence that protein constituents of postsynaptic membrane specializations are locally synthesized: analysis of proteins synthesized within synaptosomes. J. Neurosci. 11:2881–2895.

Sagara, Y. & Inesi, G. (1991). Inhibition of the sarcoplasmic reticulum Ca^{2+} transport ATPase by thapsigargin at subnanomolar concentrations. J. Biol. Chem. 266:13503–13506.

Schweitzer, E. (1987). Coordinated release of ATP and ACh from cholinergic synaptosomes and its inhibition by calmodulin antagonists. J. Neurosci. 7:2948–2956.

Sitges, M., & Talamo, B. R. (1993). Sphingosone, W7, and trifluoperazine inhibit the elevation in cytosolic calcium induced by high K^{+} depolarization in synaptosomes. J. Neurochem. 61:443–450.

Snelling, R., & Nicholls, D. (1984). The calmodulin antagonists, trifluoperazine and R24571, depolarize the mitochondria within guinea pig cerebral cortical synaptosomes. J. Neurochem. 42:1552–1557.

Sotelo, J.R., Benech, C.R., & Kun, A. (1992). Local radiolabeling of the 68kDa neurofilament protein in rat sciatic nerves. Neurosci. Let. 144:174–176.

Toeplitz, B. K., Cohen, A. I., Funke, P. T., Parker, W. L., & Gougoutas, J. Z. (1979). Structure of ionomycin: a novel diacidic polyether antibiotic having high affinity for calcium ions. J. Am. Chem. Soc. 101:3344–3353.

Tobias, G. S., & Koenig, E. (1975a). Axonal protein synthesizing activity during the early outgrowth period following neurotomy. Exp. Neurol. 49:221–234.

Tobias, G. S., & Koenig, E. (1975b). Influence of nerve cell body and neurilemma cells on local axonal protein synthesis following neurotomy. Exp. Neurol. 49:235–245.

Wang, J.K.T., Walaas, S.I., & Greengard, P. (1988). Protein phosphorylation in nerve terminals: comparison of calcium/calmodulin-dependent and calcium/diacylglycerol-dependent systems. J. Neurosci. 8:281–288.

Young, J.Z. (1962). The retina of cephalopods and its degeneration after optic section. Phil. Trans. Roy. Soc. B. 245:281–288.

Young, J.Z. (1974). The central nervous system of Loligo. I. The optic lobe. Phil. Trans. Roy. Soc. B. 267:263–302.

INDEX